AF183501

M. A. Herman

Semiconductor Superlattices

With 59 Figures

Akademie-Verlag Berlin 1986

Author:

Marian A. Herman, Ph. D.
Institute of Physics
Polish Academy of Sciences Warsaw

Erschienen im Akademie-Verlag Berlin, DDR-1086 Berlin, Leipziger Str. 3—4
© Akademie-Verlag Berlin 1986
Lizenznummer: 202 · 100/429/86
Printed in the German Democratic Republic
Gesamtherstellung: VEB Kongreß- und Werbedruck, 9273 Oberlungwitz
Lektor: Dipl-Phys. Ursula Heilmann
LSV 1185
Bestellnummer: 763 549 2 (6928)

03400

FROM THE EDITOR

The talent to synthesize "man made" semiconductor superlattices is one of the most fascinating developments of physical research for microelectronics.

The understanding of their nature is based upon lengthy and subtile lattice and electronic structure as well as lattice dynamic analyses of various kinds of superstructures especially carried out by low energy electronic diffraction (LEED) investigations.

Their preparation became reality by the development of ultra high vacuum and molecular beam epitaxy equipments as well as latterly by the very promising metalorganic chemical vapour deposition procedures which are of high interest for industrial application because of their more favourable economic parameters. MOCVD is concerned relative to their application to the most important III-V - and II-VI - compound semiconductors, MBE to Si, too.

Semiconductor superlattices have opened quite new aspects for a new quality of microelectronic circuit technology. Resulting in quantum mechanically controlled devices, e.g. in very high speed integrated (VHSI) advanced III-V compound semiconductor circuits of less than 1 fJ delay time· leakage power products or so-called InGaAsP - receivers for 1.55 μm - lasers of a $1.7 \cdot 10^9$ bit data transmission rate per second[1/] as the basis of the next generation of fibreoptic communication systems.

If superlattice silicon and semiconducting compound devices and circuits will be supplemented with appropriated one dimensional organic crystal "wires" (maybe poly-di-acetylene crystals)[2/] this could open the door to a certain kind of high speed threedimensional microelectronics, a possibly further superlattice development and application on the basis of semiconducting organic materials not included in this book but a task for a following edition when this class of material has reached a higher level of research relative to this specific application field which is only touched in the outlook chapter 6 of this book now.

Thus, this book - the first review treatment of the semiconductor superlattice topic, represents the recent state of this kind of solid-state physical research work. It is written by a competent and internationally well known solid-state physicist , affiliated since many years to the Institute of Physics of the Polish Academy of Sciences, Warsaw, Poland.

In such a way it is documented in an impressive way that high-level solid-state physical research influences more and more the progress. Seriously to inform scientists and engineers engaged in this field about urgently actual problems was the intention of writing this book.

H.G.Schneider

1/ Information from "New Scientist", May 23rd,1985 p.27

2/ Information from "New Scientist", May 30th,1985 p.23

PREFACE

Modern electronics is built upon the foundation of a well-established semiconductor technology. This technology has resulted in silicon integrated circuits, in optoelectronic devices made of III-V and II-VI semiconductor compounds, and in two-dimensional electron gas devices. It has also made feasible the realization of semiconductor superlattice structures.

The term "superlattice" is usually used to refer to a periodic structure of thin layers of two semiconductors along one dimension. The period in thickness lies typically in the range from several to tens of nanometers, which is shorter than the electron mean free path but longer than the crystal lattice constant. This periodic, or superlattice potential modifies significantly the band structure of the host semiconductors, creating minizones in wave-vector space and subbands in energy. In this regard the superlattice can be considered as a new synthesized semiconductor not present in nature, which exhibits unusual electronic and optical properties.

Essentially two types of such "man-made" superlattices may be distinguished. These are: compositional or heterojunction superlattices consisting of a periodic sequence of two semiconductors of different chemical composition, and doping superlattices, a sequence of n- and p-doped layers, possibly with undoped layers in between (nipi-crystals), grown in an otherwise homogeneous semiconductor bulk.

The superlattice potential is induced in compositional superlattices by the periodic variation of the energy gap in the direction of crystal growth, whereas in nipi-crystals it is due to the electrostatic potential of fixed ionized impurities, which may be partly compensated by mobile electrons and holes confined to the n- and p-layers, respectively.

A number of new peculiarities which are related to the electrostatic origin of the superlattice potential in nipi-crystals can be expected in addition to the phenomena, characteristic for compositional superlattices. The lowest electron and the uppermost hole subband states are locally separated from each other (an indirect semiconductor in the real space). Depending on the design of the crystal the electron-hole recombination may be suppressed nearly completely. As a consequence the effective energy gap, as well as the electron and hole concentration, can be varied in this superlattices arbitrarily over a wide range, either optically or by carrier injection into the bulk.

Today, semiconductor superlattices represent one of the fastest growing fields in solid-state physics. However, they are far more than elaborate objects of interest for physicists. Compositional superlattices, and particularly doping superlattices, with their unique tunability, represent an important new class of semiconductors, one that has quite an impact not only on solid-state physics but also on present-day electronic device technology.

The purpose of this text is to present the up-to-date knowledge on the physics of semiconductor superlattices, as well as on their practical implementation, to those who need to review or to learn superlattices. It is primarily intended for research scientists in academic and industrial laboratories. However, it will also be of use to graduate students in solid state physics or in device electronics and for persons wishing to get a general feeling of what goes on in the field. The reader is expected to be familiar with the fundamentals of solid-state physics and quantum mechanics, as well as with the basic knowledge of the physics and technology of semiconductor devices.

The text is organized into six chapters. Chapter 1 presents a general introduction to the subject and gives the outline of the variety of superlattices. Chapter 2 discusses some topics related to the epitaxial growth techniques used in the technology of superlattices. The main physical and practical aspects of molecular beam epitaxy (MBE) are presented first and then the metalorganic vapour phase epitaxy (MO VPE) is discussed. This chapter is concluded with a survey of growth peculiarities in relation to different types of superlattices.

Chapter 3 presents the most frequently used theoretical approximations which enable a theoretical description of the physical phenomena, characteristic for superlattices. Beginning with a general survey of the theory relevant to semiconductor superlattices the single quantum well theory and the effective mass approximation in relation to the type I superlattices are presented subsequently. Then the theory of type II superlattices is reviewed, including the envelope-function approximation of Bastard and the reduced Hamiltonian method of Schulman and Chang (formulated within the framework of the empirical tight-binding model). The nipi-crystal theory by Döhler and Ruden concludes this chapter.

Chapter 4 gives an up-to-date review of experimental data relevant to all types of superlattices discussed in the text. On the basis of this review in Chapter 5 the application possibilities of superlattices in different semiconductor devices are surveyed.

Chapter 6 which concludes the text, presents some trends in research and in technology of compositional and doping superlattices.

Great care has been taken to ensure that a comprehensive list of references is appended to each chapter. Therefore, this book can also serve as a bibliographical guide related to semiconductor quantum well structures and superlattices. The references to the literature are related exclusively to this chapter in which they occur. The symbols and acronyms are explained when they are used in the text for the first time. In addition the major symbols used are listed at the end of the book.

The idea of writing this text is due to H.G.Schneider, the Editor of this book. He encouraged me to work on the text and also argued for editing it by the Akademie Verlag, Berlin. I would like to express my deep gratitude to him for his manifold assistance and guidance during the publication phase of the text. Mrs. U.Heilmann, the publisher's reader, is gratefully acknowledged for her helpful advices related to the editorial side of the book.

My sincere appreciation is extended to Krzysztof Marek Herman, my son, who made all of the drawings which illustrate the text, and to Mrs. Maryla Borejko for her skillful typing of the final version of the manuscript. Finally I would like to express my thanks to Mrs. Alicja Kalińska, M.A., for the help with the editorial adjustment of the English language.

Warsaw, September, 1985

Marian A.Herman

CONTENTS

Chapter 1. STRUCTURES OF SEMICONDUCTOR SUPERLATTICES

Interest in semiconductor superlattices originates in 1970 with the proposal of Esaki and Tsu |1| to produce a one-dimensional periodic potential by a periodic variation of either impurities or alloy composition in semiconductor, with the period shorter than the electron mean free path. These authors have emphasized the compositional superlattice and have envisioned that unusual transport phenomena should occur in such "man-made" periodic structures.

Two years later Döhler |2| analysed in great detail the electronic properties of doping superlattices, and predicted a number of new peculiarities which are specific for these, so-called nipi-crystals.

Since that time, in many laboratories considerable attention has been given to searching for novel periodic structures. This has resulted in the present-day variety of semiconductor superlattices. Obviously, the intriguing physics involved in such materials has provided fuel for this advancement. Efforts in this direction apparently have opened up a new area of interdisciplinary investigations in the fields of materials science and device physics.

1.1. Historical considerations

The first artificial semiconductor superlattice was grown by molecular beam epitaxy (MBE), a thin-film growth technique developed by Cho and Arthur in the late 1960's |3|. In this technique, crystals are grown in an ultrahigh vacuum environment by directing beams of their constituent atoms or molecules from effusion cells held at elevated temperatures onto the surface of a heated single crystal substrate wafer. The beams deposit matter at a rate of about one atomic layer per second. The crystalline layers grow in this process through chemical reactions of the beams with the substrate surface.

After several attempts with different materials and processes, Chang and Esaki with their co-workers |4| succeeded in growing by MBE a compositional superlattice made of the GaAs-Al_xGa_{1-x} material system. This was not accidental. Firstly, the addition of

Al to the GaAs lattice causes little disturbance in the crystal structure because it has similar chemical valence and ion size as Ga, yet it introduces an adequate amplitude of the superlattice potential in the conduction band where the dynamics of electrons can be conveniently investigated. Secondly, the MBE technique, which was at that time the only growth technique suitable for satisfying the stringent requirements of a superlattice, has been established for these semiconductors.

The prototype structure consisted of several hundred alternating layers of undoped GaAs and $Al_xGa_{1-x}As$ with x ranging from 0.15 to 0.35 and spatial periods ranging from 5 to 20 nm in different samples. The early experiments on $GaAs/Al_xGa_{1-x}As$ superlattices were used to investigate the carriers transport anomalies |5| and to demonstrate the quantum-mechanical behaviour of electrons in rectangular quantum wells created by the band edges in these superlattices |6|.

Significant progress since then included observation of resonant Raman scattering |7| and Shubnikov-DeHaas oscillations demonstrating the two-dimensional nature of the subbands |8|, the achievement of superlattice structures with extremely thin layers |9|, and the incorporation of a superlattice region in an injection laser |10|.

The introduction of modulation doping in $GaAs/Al_xGa_{1-x}As$ superlattices |11| subsequently modified the established rectangular-well concept, and extended the underlying physics to totally new fundamental phenomena including quantized Hall effect |12|, near-zero-resistance state |13|, and electron localization in a two-dimensional electron gas (2DEG) in strong magnetic fields |14|.

The recognition of the rather unique band-edge relationship at the interface of InAs and GaSb |15| has led to a new type of compositional superlattices |16|, exemplified by the material combination of $Ga_xIn_{1-x}As$ and $GaAs_ySb_{1-y}$ |17|. For the binary compounds whose lattice constants are very close to each other, there exists the unusual case of band-edge separation: the valence band-edge of GaSb lying above the conduction band-edge of InAs (in the electronic energy space). By controlling the compositions of the ternary alloys $Ga_xIn_{1-x}As$ and $GaAs_ySb_{1-y}$ independently, this band-edge relation can be altered from separation to overlap. In addition the lattices of these alloys can be exactly matched. These features have led to new characteristics of the superlattices, beyond the scope offered by the GaAs/$Al_xGa_{1-x}As$ system. The co-existence of conduction and valence band states in the respective constituent layers causes an electron transfer from the $GaAs_ySb_{1-y}$ layers into the $Ga_xIn_{1-x}As$ layers, if the layers are sufficiently thin. This results in a semiconductor-semimetal transition |18|,|19| of the superlattice which can, in turn, be reverted back to the semiconductor state with the application of a magnetic field |20|.

A significant improvement in the spatial control of dopant incorporation during MBE growth in the direction of layer sequence down to an atomic scale was required, before the first doping superlattices could be realized by Ploog et al. |21|. The prototype of this new artificial semiconductor structure consisted of thin ($5 < d < 300$ nm) n (Si) - and p(Be) - doped GaAs layers, in some cases separated by thin intrinsic (i-GaAs) layers. Many of the electronic peculiarities of GaAs doping superlattices predicted by Döhler |2| have been then demonstrated by detailed experiments |22|.

A completely novel type of superlattices, a "spin superlattice" based on semimagnetic semiconductors |23| has been proposed by von Ortenberg |24|. The effective superlattice potential which is approximately of the square-well type, can be tuned from zero to a saturation value depending on the external magnetic field, the temperature, and the "spin doping" of the material.

Spin doping is easily achieved by substitution of only a very small amount of crystal constituents by paramagnetic ions. Thus a spin superlattice is formed by periodical substitution of paramagnetic ions into the lattice of the host crystal, so that only the effective g factor of the conduction electrons is affected. As a prototype material system for these spin superlattices $Hg_{0.99}Mn_{0.01}Se$ - $Hg_{0.976}Cd_{0.024}Se$ has been proposed. Mn serves here as the paramagnetic element, or the "spin dopant" and the concentration of the nonmagnetic constituent Cd is chosen in such a way that the energy gaps of both materials are the same.

The first spin superlattices, which are also called diluted magnetic semiconductor superlattices |25|, have been grown with the

$Cd_{1-x}Mn_xTe$ - $Cd_{1-y}Mn_yTe$ material system |26|,|27|. These superlattices have exhibited even more interesting properties than what was expected according to the predictions of paper |24|.

A new dimension has been added to the superlattice structures discussed so far by introducing a third constituent to the two host semiconductors of the superlattice, e.g., AlSb to the InAs-GaSb system |28|. These triple-constituent systems, called also polytype superlattices exhibit some characteristics which can never be achieved with the dual-constituent systems. It is therefore obvious that they extend significantly the application perspectives of man-made superlattices.

The already known variety of semiconductor superlattices calls for some classification scheme. Such a scheme should stress the characteristic features of definite superlattices and should give an insight into the materials selection problem when a desired superlattice structure has to be produced.

An attempt to construct such a classification scheme will be presented in the next section. We will begin with compositional superlattices, followed by doping superlattices, and will conclude with mixed superlattices, i.e., with doped compositional superlattices.

1.2. Compositional superlattices

The localization of the energy bandedges of different semiconductors in the electronic energy space are most often compared by using as a reference level the vacuum level |29|. Such a comparison can be done by introducing the electron affinity θ as a characteristic parameter related to each of the semiconductor materials considered.

Electron affinity specifies the energy required to release an electron from the bottom of the conduction band of the semiconductor to the vacuum level. Thus, when a semiconductor is characterized by a larger value of θ, its conduction band-edge is localized in lower position in the electronic energy space, in comparison to the conduction band-edge of a semiconductor which is characterized by a smaller value of θ. Using the vacuum level as a reference level one can divide the compositional superlattices into three different types (Fig. 1.1)

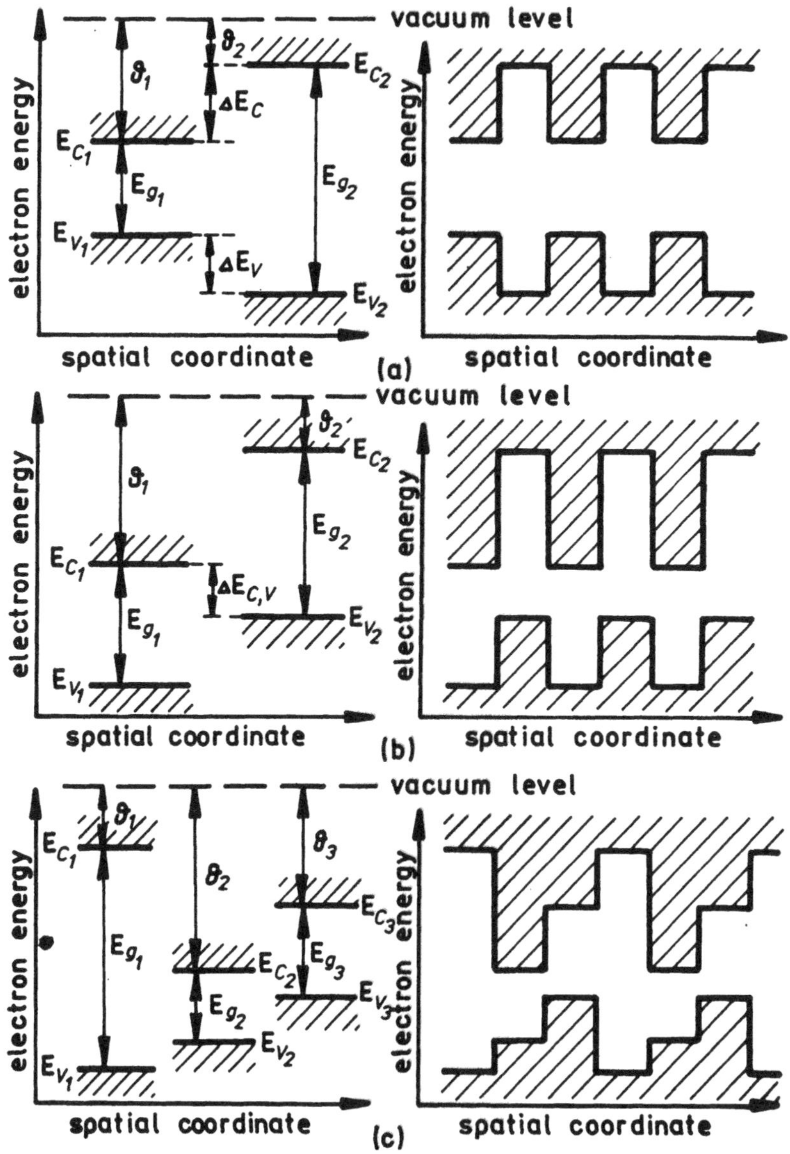

Fig. 1.1. Localization of the conduction and valence band-edges in relation to the vacuum level (left side) for the constituent compounds (not in touch) and for different types of compositional superlattices (right side).
a. type I superlattice, b. type II superlattice,
c. polytype superlattice

In the type I superlattice (Fig. 1.1a) the conduction and valence band-edge discontinuities ΔE_C and ΔE_V have opposite signs, and the energy gaps E_{g_i} overlap completely. Such superlattices are sometimes called "contra-variant" compositional superlattices |30|.

In the type II superlattice (Fig. 1.1b) the band-edge modulation has the same sign for conduction and valence bands and the energy gaps are separated or overlap only partly (this is a "co-variant" compositional superlattice |30|).

Finally, the polytype superlattice (Fig. 1.1c) is a triple-constituent material system which in addition to the constituents of the type II superlattice introduces a wide energy gap semiconductor serving as a potential barrier in this superlattice for electrons and for holes.

1.2.1. The type I compositional superlattices

The first compositional superlattice of this type has been grown of the material system GaAs-$Al_xGa_{1-x}As$. The layer sequence and the real-space energy band profile of this superlattice are illustrated in Fig. 1.2. The lineup of the bands at the heterointerfaces determines the potential barriers for the electrons and holes, and thus defines the periodic superlattice potential in the conduction and in the valence bands. The characteristic feature of this superlattice is that the narrow-gap semiconductor sandwiched between two wide-gap semiconductors forms two square quantum wells, one for electrons and one for holes. The depths of these potential wells depend on how the energy gap difference $\Delta E_g = E_{g_{AlGaAs}} - E_{g_{GaAs}}$ is divided into the band-edge discontinuities ΔE_C and ΔE_V.

Earlier papers on various aspects of square GaAs quantum wells |6| obtained experimental results consistent with the values $\Delta E_C = (0.85 \pm 0.03)\Delta E_g$ and $\Delta E_V = (0.15 \pm 0.03)\Delta E_g$. However, investigations related to parabolic quantum wells grown with the GaAs-$Al_xGa_{1-x}As$ material system |31| presented evidence for a nearly equal division of the energy gap discontinuity ΔE_g between the conduction and valence bands. This discrepancy has been removed by arguments presented in paper |32|, according to which the experimental data obtained for both the square and the parabolic GaAs quantum wells coincide when the conduction band-edge discon-

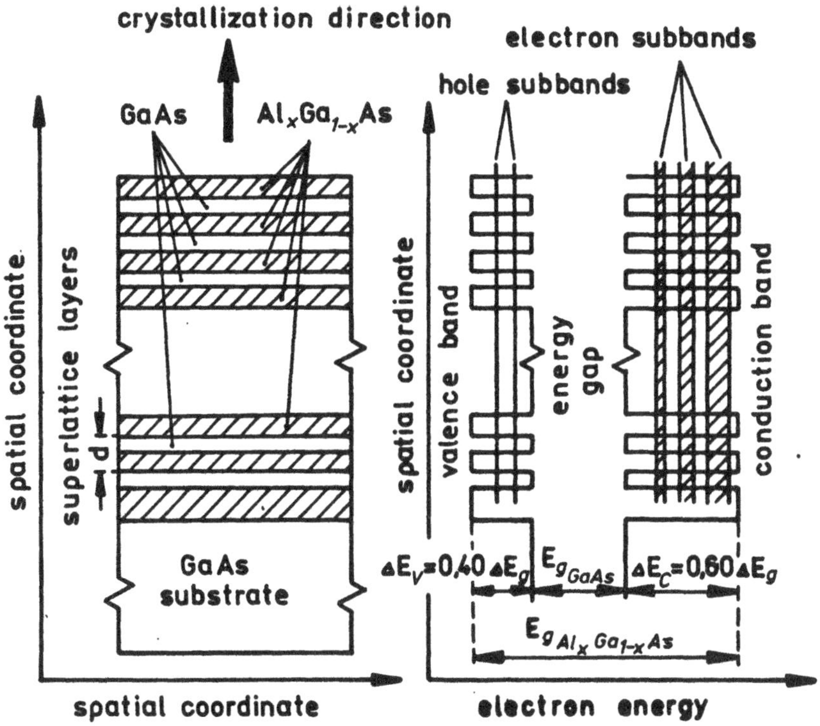

Fig. 1.2. Schematic illustration of layer sequence (left side) and of real-space energy band profile (right side) of the GaAs-$Al_xGa_{1-x}As$ superlattice

tinuity is equal to $\Delta E_C = 0.57\Delta E_g$ ($\Delta E_V = 0.43\Delta E_g$) and the following effective masses for the carriers trapped in the wells are used:

$m_e^* = 0.0665\ m_o$ for electrons,
$m_{hh}^* = 0.34\ m_o$ for heavy holes, and
$m_{lh}^* = 0.094\ m_o$ for light holes.

These values of band-edge discontinuities have been later confirmed by current transport measurements performed in an independent experimental procedure |33|. According to other experimental results |34-39| the most trustworthy values of the GaAs-$Al_xGa_{1-x}As$ heterojunction band offsets are equal to $\Delta E_C = 0.60\Delta E_g$ and $\Delta E_V = 0.40\Delta E_g$, independent of Al mole fraction in the ternary compound.

Another class of type I superlattices form the so-called strained-layer superlattices (SLS) |40-42|. These are high-quality superlattices grown from lattice-mismatched materials. The layers in these structures are kept sufficiently thin to ensure that the lattice mismatch is totally accommodated by layer strains, so that misfit defects are not generated.

The properties of SLSs can be engineered by a proper choice of the materials and geometrical parameters. For lattice mismatch of a few percent and film thickness of the order of 10 nm, good epitaxial layers can be obtained |43-45|. The lattice parameter in the directions parallel to the interfaces is then constant in the whole superlattice. Mismatch between the superlattice taken as a whole and the substrate can also be accommodated by elastic strain as long as the thickness of the superlattice stays below a critical value |44|.

Five materials systems have been chosen, so far, for growing SLSs. These are: GaAs-$In_xGa_{1-x}As$ |44|,|46|, GaAs-$GaAs_xP_{1-x}$ |40|,|47|, GaP-$GaAs_xP_{1-x}$ |48|, ZnS-ZnSe |49| and GaSb-AlSb |50|. However, it should be noticed, that new SLS structures can be grown from a wide variety of lattice-mismatched materials.

Among type I superlattices also the semimetallic-semiconductor superlattices exhibit interesting features. A typical representative of such superlattices is the HgTe-CdTe superlattice.

This superlattice presents a limiting case of a multiheterostructure of $Hg_{1-x}Cd_xTe$-CdTe, with x = 0 |54|. Typically it consists of distinct, 18 nm thick layers of HgTe (a negative-gap semimetallic compound) and 4 nm thick layers of CdTe (a wide-gap semiconductor with an energy gap of 1.49 eV at 300 K) in alternation. According to the "common anion rule" |55|, which states that the ionization potentials of two compound semiconductors are dependent on the anion species alone, the valence band-edge discontinuity in the HgTe-CdTe superlattice is nearly equal to zero. The whole energy gap difference ΔE_g, which is equal to the energy gap of CdTe is thus located solely in the conduction band in this superlattice. This leads to the rather unique relation for the conduction band discontinuity $\Delta E_C = \Delta E_g = E_{g_{CdTe}}$

The family of type I superlattices is completed by the silicon based superlattices. Within the confines of silicon materials technology (i.e., excluding III-V, II-VI and IV-VI semiconducting compounds) the number of feasible superlattice structures is greatly limited.

With regard to conventional layered structures, the Si-$Si_{1-x}Ge_x$ system is the prototype for silicon superlattices |56|. Beyond the Si-$Si_{1-x}Ge_x$ crystalline system silicon materials technology offers some other novel possibilities. One is an amorphous superlattice made of hydrogenated amorphous silicon (a-Si:H) and hydrogenated amorphous germanium (a-Ge:H) |or silicon nitride (a-SiN_x:H), or silicon carbide (a-$Si_{1-x}C_x$:H)| that are neither lattice matched nor epitaxial, yet with interfaces that are essentially defect free and nearly atomically sharp |57|,|58|. The second is an MOS (metal-oxide-silicon) configuration with a grided metal electrode made up of fine parallel metal lines uniformly spaced on the oxide layer |56|,|59|. It has been envisioned that in this system an extra periodic electron potential could be extended into the silicon by holding alternate metal lines in the grid at different voltages. The third possible silicon superlattice structure is the doping superlattice |2| which, however, will not be discussed here.

1.2.2. The type II compositional superlattices

The characteristic features of the type II compositional superlattices result from the band-edge relationship at the interfaces of the host semiconductors. The conduction band of one material is close to the valence band of the other. This causes a spatial separation of the carriers confined in the quantum wells. Electrons are confined in quantum wells existing in the first semiconductor, while holes are localized in the wells in the second semiconductor. Thus, we are dealing in these multilayer structures with an "indirect band gap in the real space" (this can be compared with the "indirect band gap in momentum space" met in many bulk semiconductors).

So far only one materials system has proven beyond doubt to create type II superlattices. This system is based on the ternaries $In_{1-x}Ga_xAs$ and $GaSb_{1-y}As_y$, with the limitting case of the InAs-GaSb superlattice for $x = 0$ and $y = 0$ |15|.

According to the available data on the electron affinity and the energy gap for the two binary materials, the conduction band-edge

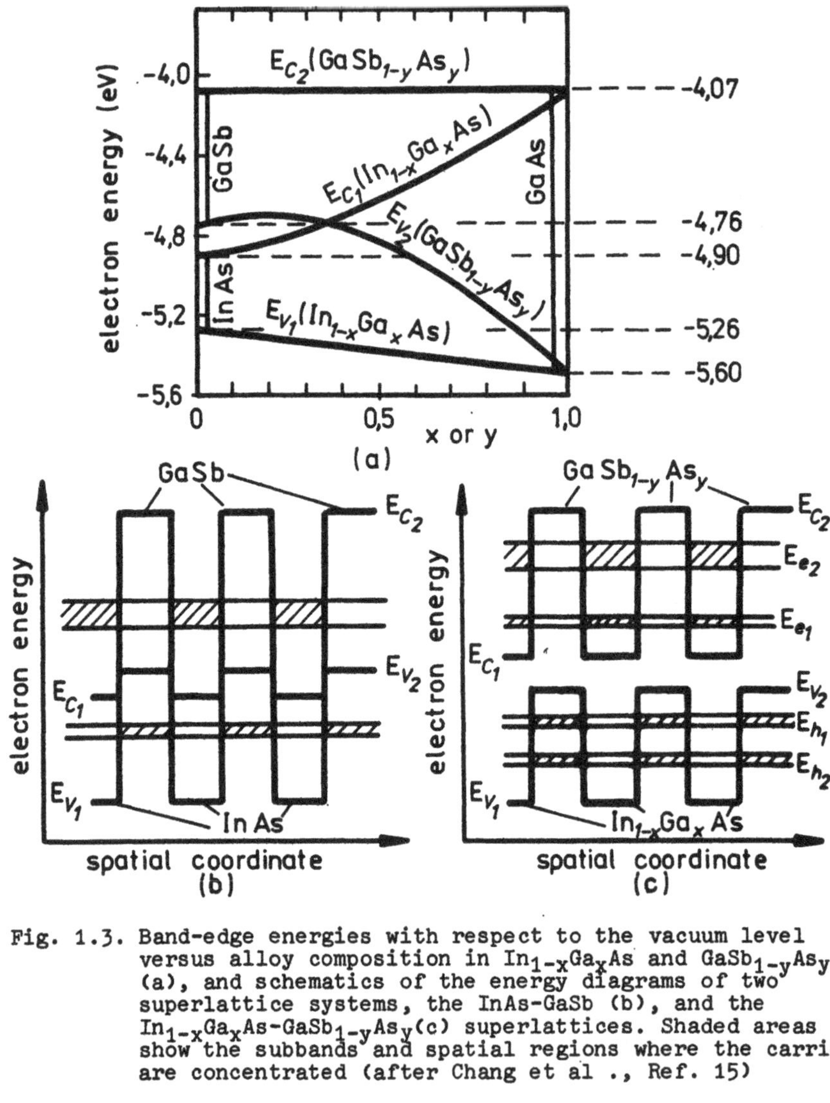

Fig. 1.3. Band-edge energies with respect to the vacuum level versus alloy composition in $In_{1-x}Ga_xAs$ and $GaSb_{1-y}As_y$ (a), and schematics of the energy diagrams of two superlattice systems, the InAs-GaSb (b), and the $In_{1-x}Ga_xAs$-$GaSb_{1-y}As_y$(c) superlattices. Shaded areas show the subbands and spatial regions where the carriers are concentrated (after Chang et al ., Ref. 15)

in InAs is at 0.14 eV below the valence band-edge in GaSb |16|. For the two ternaries the energy gaps partly overlap, with the limitting case of a complete overlap for x = 1 and y = 1 (this is the GaAs energy gap).

If the ternaries $In_{1-x}Ga_xAs$ and $GaSb_{1-y}As_y$ are selected in such a way that y = 0.918 x + 0.082, then a perfect lattice match is obtained at the hetero-interfaces of the superlattice |16|. The band-edge energies versus compound compositions as well as the schematic energy diagrams for the type II superlattices made of this materials system are shown in Fig. 1.3.

Experimental results presented recently by Murase et al. |60| raised the question of another materials system which is believed to create type II superlattices. This is the $PbTe-Pb_{1-x}Sn_xTe$ system. The electronic band structure and the type of the superlattice made of $PbTe-Pb_{1-x}Sn_xTe$ are still not understood sufficiently |61-63|, although the energy schemes in this superlattice, based on the x (tin mole fraction) dependence of the band-edge energies relative to the Fermi energy in indium doped (~1%) bulk $Pb_{1-x}Sn_xTe$ (the Fermi level is there pinned by the amphoteric In impurities) suggest |60| that superlattices made of these compounds belong to the type II family (see Fig. 1.4). This is, however, in clear contradiction to the experimental data based on magnetooptical experiments performed on this material system, because according to these data |61|,|63| the $PbTe-Pb_{1-x}Sn_xTe$ superlattices belong to the type I family.

1.2.3. Polytype superlattices

As discussed earlier, polytype superlattices can be obtained by addition of AlSb to the InAs-GaSb material system (see, Fig.1.1c). The band-edge energies of AlSb relative to those of GaSb and InAs are illustrated in Fig. 1.5a. The minimum of the conduction band for AlSb is located at the X point in the momentum space, although those for GaSb and InAs are at the Γ point. Nevertheless, since the relatively wide energy gap of AlSb (1.6 eV) extends over those of GaSb and InAs,AlSb layers serve as potential barriers for electrons as well as for holes in superlattices and heterostructures. It should be realized that the lattice constant of AlSb (6.136 Å) is compatible with those of GaSb (6.095 Å) and InAs (6.058 Å) |28| for heteroepitaxy. Therefore, these semiconductors, significantly differing in band parameters, represent a rather unique combination among III-V compound semiconductors.

The basic elements involved in such polytype superlattices are: BAC, ABCA, ACBCA,...multi-heterojunctions (Fig. 1.5b) with A standing for AlSb,B for GaSb and C for InAs.

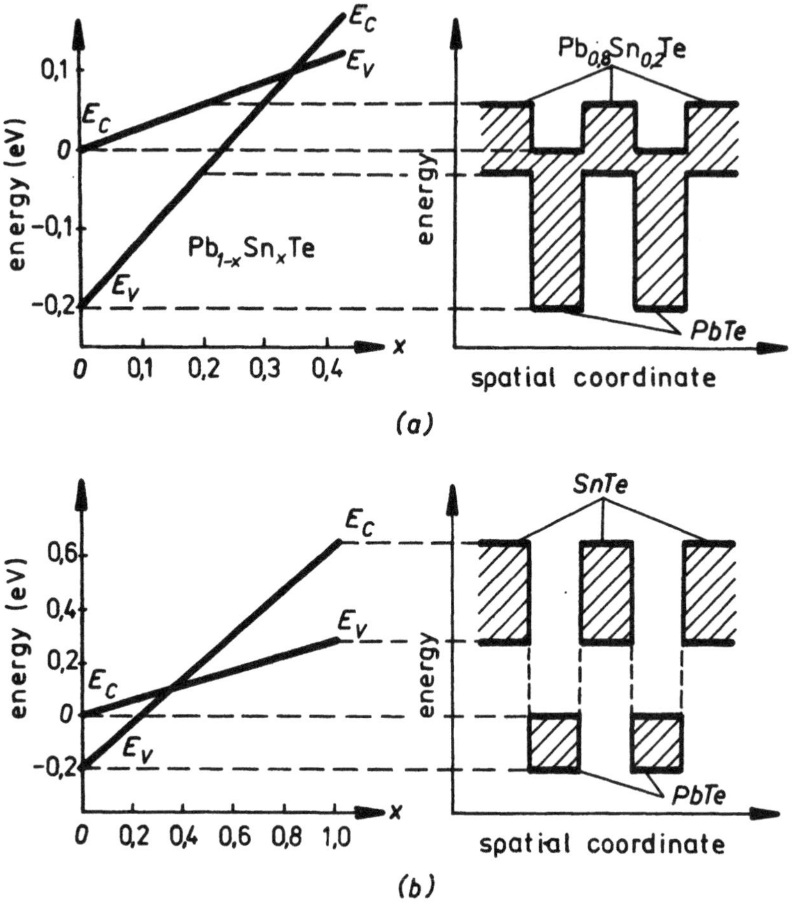

Fig. 1.4. Tin composition dependent band-edge energies in $Pb_{1-x}Sn_xTe$ deduced by taking the Fermi level pinned by amphoteric indium impurities (~1%) as energy reference level (left side), and the schematics of the energy diagrams of two superlattice systems, the $PbTe-Pb_{0.8}Sn_{0.2}Te$ (a), and the PbTe-SnTe (b) superlattices (after Murase et al.,Ref. 60).

Heterojunctions between the constituent compounds of a polytype superlattice show interesting characteristics in their own right |28|. The GaSb-AlSb heterojunctions exhibit, for example, very interesting energy band features. According to the common anion rule |55|,|64| the valence band offset ΔE_V should be very small, however, the conduction band-edge discontinuity ΔE_C is very large.

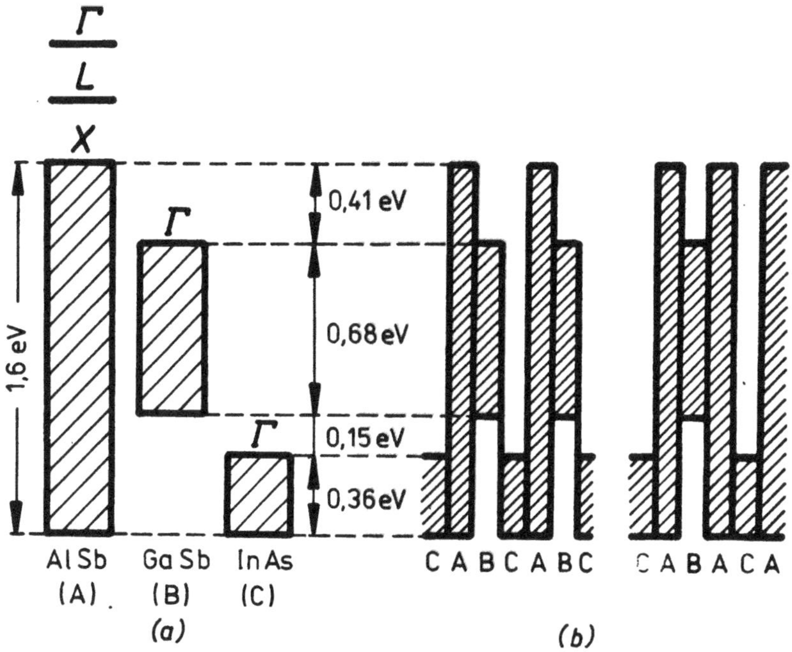

Fig. 1.5. a) Estimated band-edge energies of AlSb relative to those of GaSb and InAs (the hatched areas indicate the energy gaps), and b) energy diagrams of two kinds of polytype superlattices (after Esaki et al., Ref. 28).

Consequently, quantum wells in the conduction band of a GaSb-AlSb superlattice will be so deep that the energies of the quantized levels will not depend on the exact value of ΔE_C. Also, as the lattice mismatch between GaSb and AlSb is not small $\Delta a_o/a_o = 0.65\%$ |50|, this system offers opportunities for the study of the effects induced by misfit strains in a superlattice .

It is obvious that all characteristics of the heterojunctions between the constituent compounds define also the properties of the polytype superlattice as the whole.

1.3. Doping superlattices

The term "doping superlattice" is used to refer to a periodic array consisting of layers of the same semiconductor doped in two different ways. In the n-doped layers donor atoms contribute electrons, while in the p-doped layers acceptor atoms bind electrons. The resulting distribution of electronic charge creates a set of parabolic potential wells.

The unusual electronic properties of doping superlattices derive from the specific nature of the superlattice potential which, in this case, is the space charge potential of ionized impurities in the doping layers |2|,|22|,|65|. This is in contrast to the compositional superlattices, in which the superlattice potential originates from the different band gap values of their composing materials. The space charge potential in the doping superlattices modulates the band-edges of the host material in such a way that electrons and holes become spatially separated. This separation can be made nearly perfect by the appropriate choice of the design parameters, which are the doping concentrations and the layer thicknesses.

One of the attractive features of doping superlattices is that any semiconductor that can be doped both n- and p-type in a well controlled way can be used as the host material. GaAs grown by MBE is used most frequently |65|, but also Si |56|, as well as PbTe, $Pb_{1-x}Ge_xTe$ or $Pb_{1-x}Sn_xTe$ compounds have been proposed |66|. Another particular advantage of doping superlattices is related to the structural perfection of this multilayer system. The relatively small amounts of impurities used in doping superlattices, typically 10^{17} - 10^{19} cm^{-3} |65|, induce only a minor distortion of the lattice of the host material. Thus, doping superlattices do not contain any of the typical material interfaces of the compositional superlattices with the possibility of alloy disorder, or misfit strains. We are dealing with a homogeneous semiconductor which is only modulated by the alternating character of doping. Moreover, the effective band gap of a doping superlattice can be given any value from zero to the band gap characteristic of the undoped host material by choosing the appropriate combination of doping concentration and layer thickness.

Let us assume a semiconductor with a band gap E_g, with a static dielectric constant ε, which is modulated by periodic n- and p-doping in the z-direction. Although the doping profiles $N_D(z)$ and $N_A(z)$ may have any arbitrary shape, for the sake of simplicity only the simplest case of constant and equal doping concentrations $N_A = N_D$ within the n- and p-doped regions of thicknesses $d_n = d_p = d/2$ will be considered here. Disregarded here are also the possibility of impurity band formation and the possible random inhomogeneities in the distribution of the impurity atoms.

In the macroscopically compensated situation described, all impurities will be ionized in the ground state due to recombination of the electrons from donors with the holes on the acceptors. Therefore, a periodic parabolic space-charge (or built-in) potential $V_D(z)$ due to the donors (positively charged) and the acceptors (negatively charged) exists in the crystal. This potential, having an amplitude |65| (in the cgs units system)

$$V_{D_o} = (2\pi e^2 N_D/\varepsilon)(d/2)^2 \tag{1.3.1}$$

is superimposed to the crystal lattice potential, and modulates the conduction and valence band-edges as shown schematically in Fig. 1.6.

The motion of charge carriers in the z-direction is then quantized by the superlattice potential $V_D(z)$. The effective band gap $E_{g_{SL}}$ of the superlattice, i.e., the distance between the bottom of the lowest conduction subband at the energy $E_{C,O}$ above the conduction band minimum and the top of the uppermost valence subband at $E_{V,O}$ below the valence band maximum becomes

$$E_{g_{SL}} = E_g - 2V_{D_o} + E_{C,O} + |E_{V,O}| \tag{1.3.2}$$

It is obvious that the effective band gap $E_{g_{SL}}$ is strongly reduced below the value E_g of the unmodulated host material, and that it depends on V_{D_o}, that is on doping concentrations and the layer thicknesses.

1.4. Doped compositional superlattices

In standard compositional superlattices the layers of the constituent semiconductors are undoped. Therefore, the quantum wells can be treated as square-shaped (see, Fig. 1.2). The situation will be changed if one of the semiconductors of the superlattice, e.g., the wide-gap semiconductor will be doped with donors. This special situation has been demonstrated in the GaAs-$Al_xGa_{1-x}As$ modulation doped superlattice |11|. Since the GaAs conduction

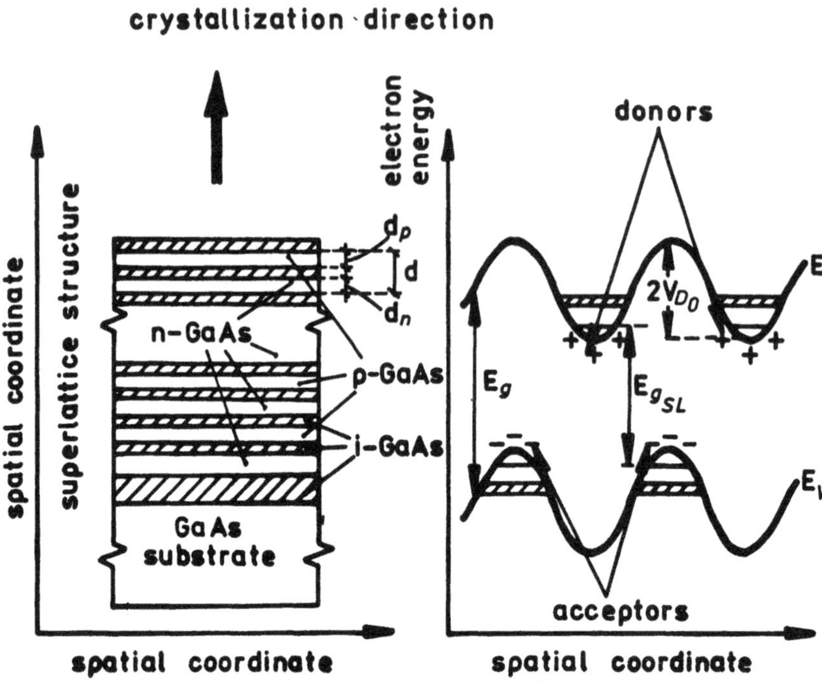

Fig. 1.6. Schematic illustration of layer sequence (left side) and of real-space energy band profile (right side) of the GaAs doping superlattice (GaAs nipi-crystal)

band-edges lie lower in energy than the $Al_xGa_{1-x}As$ donor states do. electrons released from the donors into the $Al_xGa_{1-x}As$ crystal lattice can diffuse toward the undoped GaAs regions gaining thus spatial separation from their parent ionized donors. Finally, all mobile carriers become confined to the GaAs layers where they can move in the directions parallel to the hetero-interface without being scattered by ionized impurities. This leads to considerable enhancement of the mobility of these electrons in the narrow-gap semiconductor layers.

The spatial transfer of mobile charge carriers create, however, also space charge regions of alternating sign in the superlattice, what results in a periodic bending of the band-edges, as shown in Fig. 1.7. The electrons, which occupy the quantum well states in

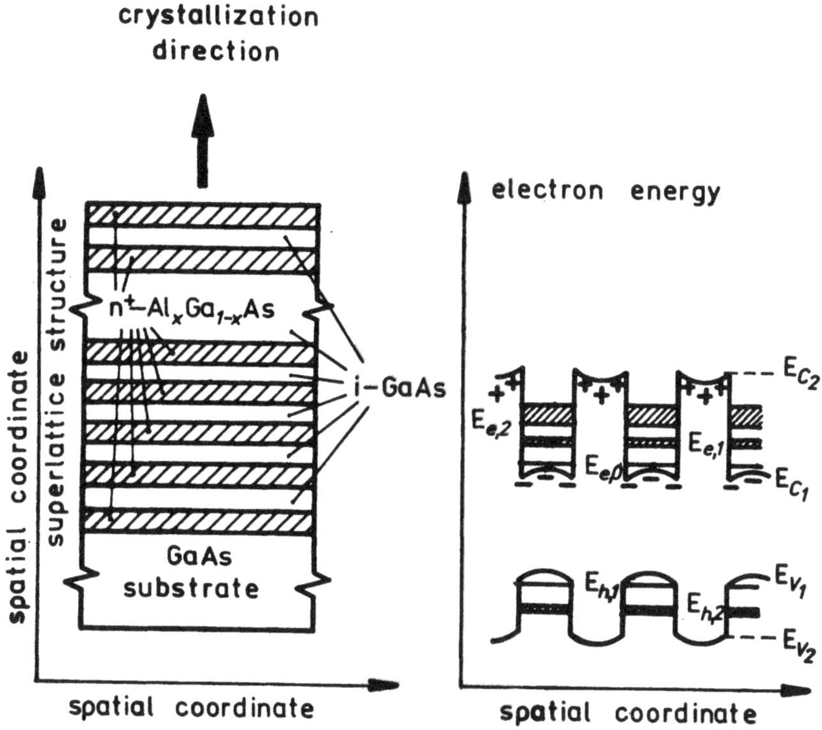

Fig. 1.7. Schematic illustration of layer sequence (left side) and of corresponding real-space energy band diagram (right side) of the i-GaAs - n^+-$Al_xGa_{1-x}As$ modulation doped superlattice. The band bending at the constituent hetero-interfaces arises from the space charges, due to a spatial transfer of free electrons from the ionized donors in the n^+-$Al_xGa_{1-x}As$ barriers into the i-GaAs wells.

GaAs up to the Fermi level E_F exhibit higher mobility enhancement in modulation doped superlattices in which thin undoped wide-gap spacer layers were inserted |67|, i.e., a barrier regions of 5-10 nm thickness near the GaAs interfaces were left undoped. The effect of mobility enhancement is most pronounced in these superlattices at low temperatures where the phonon scattering processes competing with the Coulomb scattering processes (largely diminished) are weak. Peak mobilities in excess of 10^6 $cm^2/(V \cdot s)$, at 4.2 K have been achieved |68|.

As can be seen from Fig. 1.7 the established square-well concept for the type I superlattices has to be modified toward parabolic like quantum wells in modulation doped superlattices.

A new heterojunction doping superlattice exhibiting tunability of the electronic properties (a feature characteristic for doping superlattices) and simultaneously significant mobility enhancement of both, electrons and holes confined in quantum wells (a feature characteristic for modulation doped superlattices), has been proposed by Döhler |69| and realized using MBE growth technique by Künzel and co-workers |70|. In this superlattice a periodic sequence of ten individual ultrathin layers differently doped is applied. The structure consists in general of a periodic sequence of n-Al_xGa_{1-x}As/i-Al_xGa_{1-x}As/i-Al_yGa_{1-y}As/i-Al_xGa_{1-x}As/n-Al_xGa_{1-x}As/p-Al_xGa_{1-x}As/i-Al_xGa_{1-x}As/i-Al_yGa_{1-y}As/i-Al_xGa_{1-x}As/p-Al_xGa_{1-x}As layered stacks with $y \ll x$, or $y = 0$. The basic idea for forming heterojunction doped superlattices of this kind was to modify a conventional doping superlattice of higher band gap material, e.g. $Al_xGa_{1-x}As$, by the periodic insertion of nominally undoped i-layers of lower band gap material, e.g. $Al_yGa_{1-y}As$. To illustrate this Fig. 1.8 shows the arrangement of the layer sequence for the periodic structure under consideration, with $y = 0$, and the energy band-edge profile (in real space) of the one period section of this structure.

Ultrathin undoped i-GaAs layers are sandwiched between alternating n- and p-doped $Al_xGa_{1-x}As$ regions. The desired effect of spatial separation between free carriers transferred to the GaAs layers and their parent ionized impurities is enhanced by inserting thin undoped i-$Al_xGa_{1-x}As$ spacer layers at the heterojunctions. Thus, the period of this superlattice consists of one n- and one p-doped $Al_xGa_{1-x}As$ region, each interspersed with three undoped layers, resulting in an overall of 10 layers per superlattice period. This means, however, that the conventional doping superlattice potential is periodically interrupted by the potential wells created by the lower band gap material regions. As a result, the motion of free carriers parallel to the constituent layers is nearly free (apart from residual impurity and phonon scattering processes) whereas carrier motion in the direction of periodicity is quantized by the one-dimensional potential modulating the band-edges.

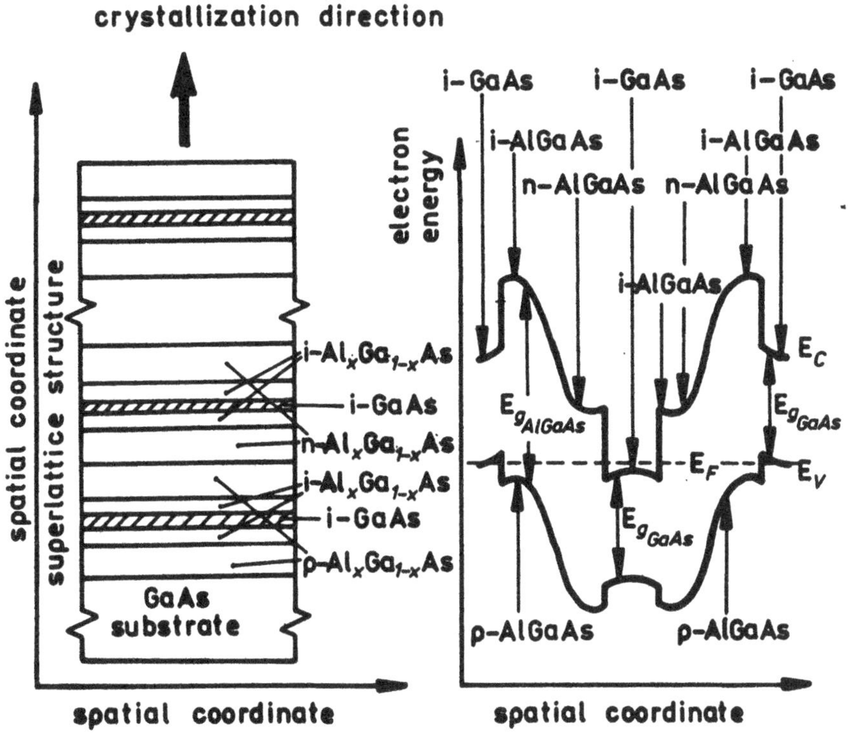

Fig. 1.8. Schematic illustration of layer sequence (left side) and energy band-edge profile in real space (right side) of the GaAs-Al_xGa_{1-x}As heterojunction doping superlattice (not drawn to scale). The superlattice period consists of as many as ten individual layers.

The amplitude of the periodic space charge potential in the ground state is determined by the doping concentration and by the thickness of the alternately doped Al_xGa_{1-x}As regions. Therefore, heterojunction doping superlattices having a wide range of tailored band profiles can be produced intentionally by preselecting their design parameters for epitaxial growth. In addition, the space charge potential and thereby the band structure may be tuned by electrical or by optical excitation of the crystal.

1.5. Classification schemes for semiconductor superlattices

Summarizing this chapter in which a survey of the present-day variety of semiconductor superlattices has been presented, the following classification schemes can be introduced.

Fig. 1.9. shows a general classification scheme based on structural parameters of the superlattices. Here the main classification criterion is related to the crystallinity of the layers creating

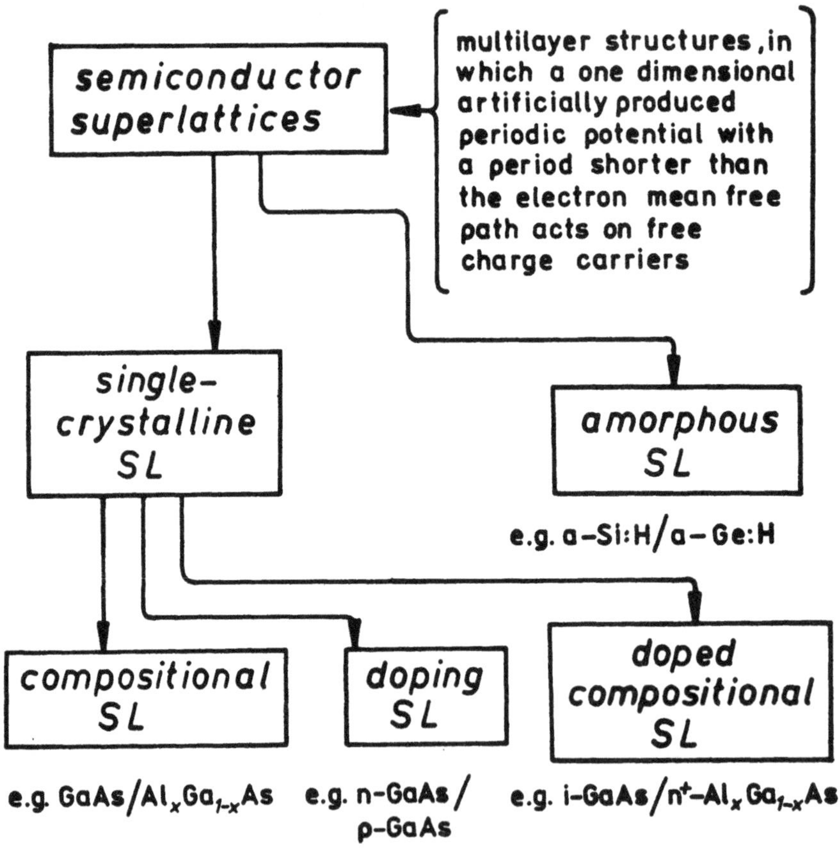

Fig. 1.9. General classification scheme of semicondictor superlattices.

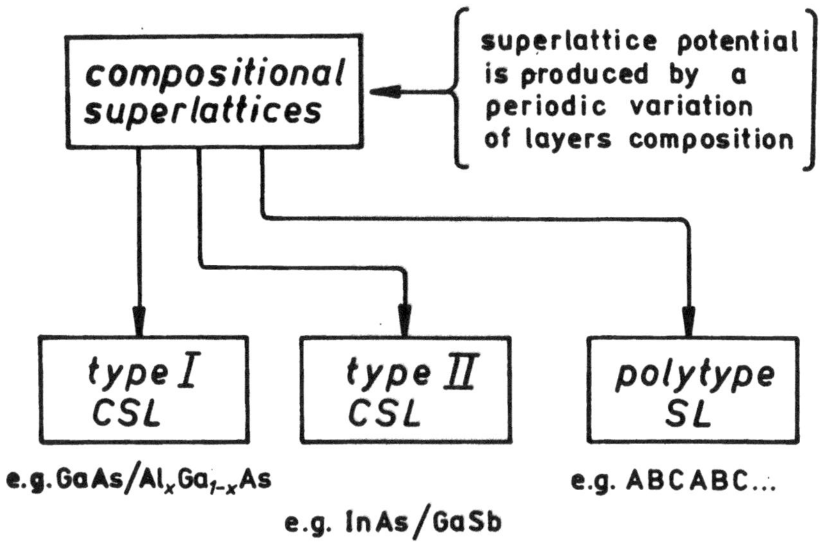

Fig. 1.10. Classification scheme of compositional superlattices

the superlattice. As the secondary criterion the physical origin of the superlattice potential is used. Fig. 1.10 shows the classification scheme of compositional superlattices. The classification criterion here is based on the characteristic lineup of the band-edges at the hetero-interfaces in the superlattice. Finally, Fig. 1.11 presents a classification scheme related to type I compositional superlattices. Here the specific materials used to produce the superlattice structure and matching of the crystal lattices at the hetero-interfaces of the superlattice have been chosen as classification criteria.

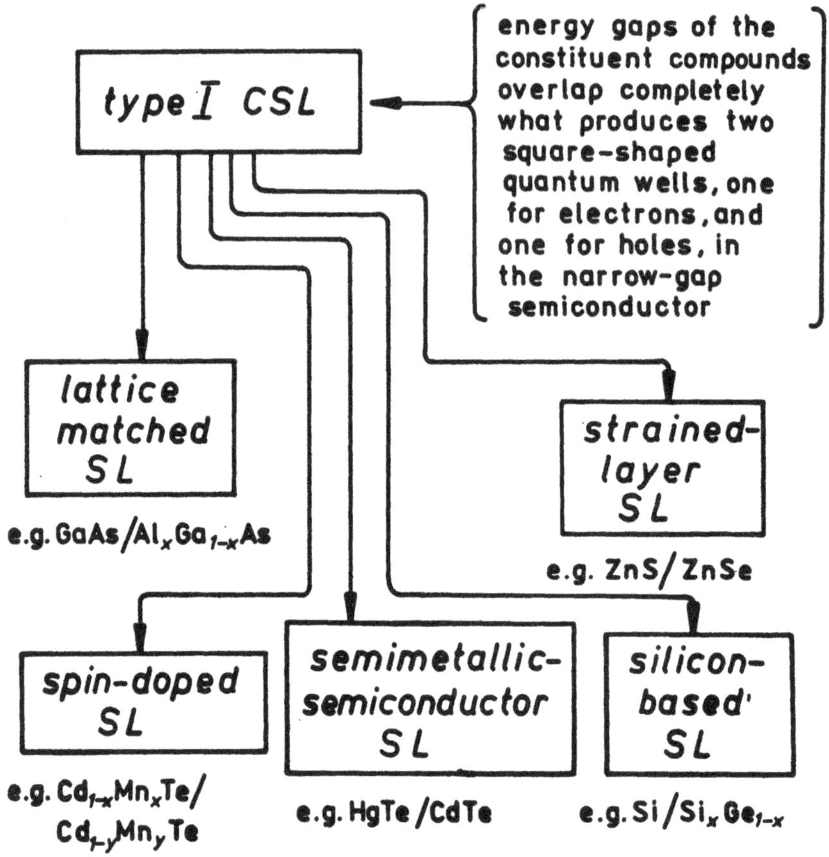

Fig. 1.11. Classification scheme of type I compositional superlattices.

1.6. Literature to chapter 1.

|1| Esaki, L., and R. Tsu: Superlattice and negative differential conductivity in semiconductors, IBM J.Res.Dev. 14 (1970) 1, 61-65

|2| Döhler, G.H.: Electron states in crystals with "nipi-superstructure", Phys.Stat.Sol. (b) 52 (1972) 1,79-92, ibid 52 (1972) 2,533-545

|3| Cho, A.Y., and J.R.Arthur: Molecular beam epitaxy, Progr.Solid State Chem. 10 (1975) 3,157-191

|4| Chang.L.L., L.Esaki, W.E.Howard, and R.Ludeke: The growth of a GaAs-AlGaAs superlattice, J.Vac.Sci.Technol. 10 (1973) 1,11-16

|5| Esaki,L, and L.L.Chang: Semiconductor superfine structures by computer-controlled molecular beam epitaxy, Thin Solid Films 36 (1976) 1,285-298

|6| Dingle,R: Confined carrier quantum states in ultrathin semiconductor heterostructures, Advances in Solid State Physics XV, 21-58, Queisser H.J. (ed), Pergamon-Vieweg, Braunschweig 1975.

|7| Manuel,P., G.A.Sai-Halasz, L.L.Chang, C.A.Chang, and L.Esaki: Resonant Raman scattering in a semiconductor superlattice, Phys.Rev.Lett. 37 (1976) 25,1701-1704

|8| Chang,L.L., H.Sakaki, C.A.Chang, and L.Esaki: Shubnikov-de Haas oscillations in a semiconductor superlattice, Phys.Rev.Lett. 38 (1977) 25, 1489-1493

|9| Gossard,A.C., P.M.Petroff, W.Wiegmann, R.Dingle, and A. Savage: Epitaxial structures with alternate-atomic-layer composition modulation, Appl.Phys.Lett. 29 (1976) 6,323-325

|10| Holonyak,Jr., N.,R.M.Kolbas, W.D.Laidig, and B.A.Vojak: Low-threshold continuous laser operation (300-337 K) of multilayer MO-CVD $Al_xGa_{1-x}As$-GaAs quantum-well heterostructures, Appl.Phys.Lett. 33 (1978) 8,737-739

|11| Dingle,R., H.L.Störmer, A.C.Gossard, and W.Wiegmann: Electron mobilities in modulation-doped semiconductor heterojunction superlattices, Appl.Phys.Lett. 33 (1978) 7,665-667

|12| von Klitzing, K., H.Obloh, G.Ebert, J.Knecht, and K.Ploog: The quantized Hall resistance in two-dimensional systems, in: Precision Measurement and Fundamental Constants II, 519-528, B.N.Taylor and W.D.Phillips (eds), Natn.Bur.Stand. (U.S.) Spec.Publ.No 617 1984.

|13| Tsui,D.C., H.L.Störmer, and A.C.Gossard: Zero-resistance state of two-dimensional electrons in a quantizing magnetic field, Phys.Rev.B 25 (1982) 2,1405-1407

|14| Tsui,D.C., H.L.Störmer, and A.C.Gossard: Two-dimensional magneto-transport in the extreme quantum limit, Phys.Rev.Lett. 48 (1982) 22,1559-1562

|15| Chang,L.L., and L.Esaki: Electronic properties of InAs-GaSb superlattices, Surface Sci. 98 (1980) 1,70-89

|16| Sai-Halasz,G.A., R.Tsu, and L.Esaki: A new semiconductor superlattice, Appl.Phys.Lett. 30 (1977) 12,651-653

|17| Sakaki,H., L.L.Chang, R.Ludeke, C.A.Chang, G.A.Sai-Halasz, and L.Esaki: $In_{1-x}Ga_xAs$-$GaSb_{1-x}As_x$ heterojunctions by molecular beam epitaxy, Appl.Phys.Lett. 31 (1977) 3,211-213

|18| Sai-Halasz,G.A., L.Esaki, and W.A.Harrison: InAs-GaSb superlattice energy structure and its semiconductor-semimetal transition, Phys.Rev.B 18 (1978) 6,2812-2818

|19| Chang,L.L., N.Kawai, G.A.Sai-Halasz, R.Ludeke, and L.Esaki: Observation of semiconductor-semimetal transition in InAs-GaSb superlattice, Appl.Phys.Lett 35 (1979) 12,939-942

|20| Kawai,N., L.L.Chang, G.A.Sai-Halasz, C.A.Chang, and L.Esaki: Magnetic field-induced semimetal-to-semiconductor transition in InAs-GaSb superlattice, Appl.Phys.Lett. 36 (1980) 5, 369-371

|21| Ploog,K., A.Fischer, and H.Künzel: The use of Si and Be impurities for novel periodic doping structures in GaAs grown by molecular beam epitaxy, J.Electrochem.Soc. 128 (1981) 2, 400-410

|22| Döhler,G.H.: nipi doping superlattices-tailored semiconductors with tunable electronic properties, Advances in Solid State Phys. XXIII, 207-226, Grosse P. (ed.), Vieweg, Braunschweig 1983.

|23| Galazka,R.R.: Semimagnetic semiconductors based on HgMnTe and CdMnTe, in: Physics of Semiconductors 1978, Wilson B.L.H. (ed.), Inst.Phys.Conf.Ser.43, Institute of Physics, London, (1979) 133-140

|24| von Ortenberg,M.: Spin superlattice with tunable minigap, Phys.Rev.Lett. 49 (1982) 14,1041-1043

|25| Datta,S., J.K.Furdyna, and R.L.Gunshar: Diluted magnetic semiconductor superlattices and heterostructures, Superlattices and Microstructures, 1 (1985) 4,327-334

|26| Kołodziejski,L.A., T.C.Bonsett, R.L.Gunshor, S.Datta, R.B. Bylsma, W.M.Becker, and N.Otsuka: MBE of diluted magnetic semiconductor $Cd_{1-x}Mn_xTe$ superlattices, Appl.Phys.Lett. 45 (1984) 4,440-442

|27| Bicknell,R.N., R.W.Yanka, N.C.Giles-Taylor, D.K.Blanks, E.L. Buckland, and J.F.Schetzina: $Cd_{1-x}Mn_xTe$-CdTe multilayers grown by MBE, Appl.Phys.Lett. 45 (1984) 1,92-94

|28| Esaki,L., L.L.Chang, and E.E.Mendez: Polytype superlattices and multi-heterojunctions, Jap.J.Appl.Phys. 20 (1981) 7, L529-L532

|29| Gobelli,G.W., and F.G.Allen: Photoelectric properties of cleaved GaAs,GaSb,InAs, and InSb surfaces, comparison with Si and Ge, Phys.Rev. 137A (1965) 1,245-254

|30| Shik,A.Ya: Superlattices-periodic semiconductor structures, Fiz.Tekh.Poluprov. 8 (1974) 10,1841-1864

|31| Miller,R.C., A.C.Gossard, D.A.Kleinman, and O.Munteanu: Parabolic quantum wells with the GaAs-$Al_xGa_{1-x}As$ system, Phys. Rev. B 29 (1984) 6,3740-3743

|32| Miller,R.C., D.A.Kleinman, and A.C.Gossard: Energy-gap discontinuities and effective masses for GaAs-$Al_xGa_{1-x}As$ quantum wells, Phys.Rev.B 29 (1984) 12,7085-7087

|33| Batey,J., S.L.Wright, and D.J.DiMaria: Energy band-gap discontinuities in GaAs/AlGaAs heterojunctions, J.Appl.Phys. 57 (1985) 2,484-487

|34| Okumura,H., S.Misawa, S.Yoshida, and S.Gonda:Determination of the conduction-band discontinuities of GaAs/$Al_xGa_{1-x}As$ interfaces by capacitance-voltage measurements, Appl.Phys. Lett. 46 (1985) 4,377-379

|35| Kroemer,H.: Determination of heterojunction band offsets by capacitance-voltage profiling through nonabrupt isotype heterojunctions, Appl.Phys.Lett. 46 (1985) 5,504-505

|36| Arnold,D., A.Ketterson, T.Henderson, J.Klem, and H.Morkoc: Electrical characterization of GaAs/AlGaAs semiconductor-insulator-semiconductor capacitors and application to the

measurement of the GaAs/AlGaAs band-gap discontinuity, J. Appl.Phys. 57 (1985) 8,2880-2885

|37| Kelly,M.K.,D.W.Niles, E.Colavita, G.Margaritondo, and M.Henzler: Valence-band discontinuities at AlAs-based heterojunction interfaces, Appl.Phys.Lett. 46 (1985) 8,768-770

|38| Hickmott,T.W., P.M.Solomon, R.Fischer, and H.Morkoc: Negative charge, barrier heights, and the conduction-band discontinuity in $Al_xGa_{1-x}As$ capacitors, J.Appl.Phys. 57 (1985) 8, 2844-2853

|39| Meynadier,M.H., C.Delalande, G.Bastard, M.Vos, F.Alexandre, and J.L.Lievin: Size quantization and band-offset determination in GaAs-AlGaAs separate confinement heterostructures, Phys.Rev. B 31 (1985) 8,5539-5542

|40| Matthews,J.W., and A.E.Blakeslee: Almost perfect epitaxial multilayers, J.Vac.Sci.Technol. 14 (1977) 4,989-991

|41| Osbourn,G.C.: Strained-layer superlattices from mismatched materials, J.Appl.Phys. 53 (1982) 3, 1586-1589

|42| Osbourn,G.C.: $In_xGa_{1-x}As$ - $In_yGa_{1-y}As$ strained-layer superlattices: A proposal for useful, new electronic materials, Phys.Rev. B 27 (1983) 8,5126-5128

|43| Dodson,B.W.: Stability of registry in strained-layer superlattice interfaces, Phys.Rev. B 30 (1984) 6,3545-3546

|44| Quillec,M., L.Goldstein, G.LeRoux, J.Burgeat, and J.Primot: Growth conditions and characterization of InGaAs/GaAs strained layer superlattices, J.Appl.Phys. 55 (1984) 8,2904-2909

|45| Tamargo,M.C., R.Hull, L.H.Greene, J.R.Hayes, and A.Y.Cho: Growth of a novel InAs-GaAs strained-layer superlattice on InP, Appl.Phys.Lett. 46 (1985) 6,569-571

|46| Brown,J.M., N.Holonyak, Jr, R.W.Kaliski, M.J.Ludowise, W.T Dietze, and C.R.Lewis: Effect of layer size on lattice distortion in strained-layer superlattices, Appl.Phys.Lett. 44 (1984) 12,1158-1160

|47| Guorley,P.L., and R.M.Biefeld: Quantum size effects in $GaAs/GaAs_xP_{1-x}$ strained-layer superlattices, Appl.Phys. Lett. 45 (1984) 7, 749-751

|48| Speriosu, V.S., M.A.Nicolet, S.T.Picraux, and R.M.Biefeld: Depth profiles of perpendicular and parallel strain in a $GaAs_xP_{1-x}/GaP$ superlattice, Appl.Phys.Lett. 45 (1984) 3, 223-225

|49| Fujiyasu,H., H.Takahashi, H.Shimizu, A.Sasaki, and H.Kuwabara: Optical properties of ZnS-ZnSe superlattices prepared by a HWE, Proceedings of the 17th Intern.Conf.on Physics of Semiconductors, San Francisco, 1984

|50| Voisin,P., C.Delalande, G.Bastard, M.Voos, L.L.Chang, A. Segmuller, C.A.Chang, and L.Esaki: Electronic structure of GaSb-AlSb superlattices, Superlattices and Microstructures, 1 (1985) 2,155-159

|51| Faurie, J.P., A.Million, and J.Piaguet: CdTe-HgTe multilayers grown by molecular beam epitaxy, Appl.Phys.Lett. 41 (1982) 8,713-715

|52| Faurie,J.P., A.Million, R.Boch, and J.L.Tissot: Latest developments in the growth of $Cd_xHg_{1-x}Te$ and CdTe-HgTe superlattices by molecular beam epitaxy, J.Vac.Sci.Technol. A1 (1983) 3,1593-1597

|53| Faurie,J.P., M.Boukerche, S.Sivananthah, J.Reno, and C.Hsu: HgTe/CdTe superlattices grown by MBE, Superlattices and Microstructures 1 (1985) 3, 237-244

|54| Herman,M.A., and M.Pessa: $Hg_{1-x}Cd_xTe$-$Hg_{1-y}Cd_yTe$ (0<x,y<1) heterostructures: Properties, epitaxy, and applications, J. Appl.Phys. 57 (1985) 8,2671-2694

|55| McCaldin,J.O., T.C.McGill, and G.A.Mead: Correlation for III-V and II-VI semiconductors of the Au Schottky barrier energy with anion electronegativity, Phys.Rev.Lett. 36 (1976) 1,56-58

|56| Krishnamurthy,S., and J.A.Moriarty: Electronic structure of silicon superlattices, Superlattices and Microstructures, 1 (1985) 3,209-215

|57| Abeles,B., and T.Tiedje: Amorphous semiconductor superlattices, Phys.Rev.Lett. 51 (1983) 21,2003-2006

|58| Abeles,B., T.Tiedje, H.C.Stasiewski, H.W.Deckman, P.D. Persans, K.S.Liang, and C.B.Roxlo: Amorphous semiconductor superlattices, Superlattices and Microstructures, 1 (1985) 2,115-118

|59| Stiles,P.J.: The artificial superlattice in a two-dimensional system, Surface Sci. 73 (1978) 1,252-257

|60| Murase,K., S.Shimomura, S.Takaoka, A.Ishida, and H.Fujiyasu: On the type of superlattices in lead-tin-telluride system, Superlattices and Microstructures, 1 (1985) 2,177-182

|61| Ambrosch,K.E., H.Clemens, E.J.Fantner, G.Bauer, M.Kriechbaum, P.Kocevar, and R.J.Nicholas: Structural and electronic properties of $PbTe/Pb_{1-x}Sn_xTe$ superlattices, Surface Sci. 142 (1984) 2,571-578

|62| Fujiyasu,H., A.Ishida, H.Kuwabara, S.Shimomura, S.Takaoka, and K.Kurase: Optical and electrical properties of $PbTe$-$Pb_{1-x}Sn_xTe$ superlattices prepared on KCl by HWE, Surface Sci. 142 (1984) 2,579-585

|63| Pichler,P., E.J.Fantner, G.Bauer, H.Clemens, H.Pascher, M.von Ortenberg, and M.Kriechbaum: Magnetooptical investigation of $PbTe/Pb_{1-x}Sn_xTe$ superlattices, Superlattices and Microstructures, 1 (1985) 1,1-9

|64| Harrison, W.A.: Elementary theory of heterojunctions, J.Vac. Sci.Technol. 14 (1977) 4,1016-1021

|65| Ploog,K., and G.H.Döhler: Compositional and doping superlattices in III-V semiconductors, Advances in Phys. 32 (1983) 3,285-359

|66| Ruden,P.P., T.L.Reinecke, and F.Crowne: Temperature dependence of electronic and lattice properties of doping superlattices in IV-VI semiconductors, Superlattices and Microstructures, 1 (1985) 3,197-200

|67| Störmer,H.L., A.Pinczuk, A.C.Gossard, and W.Wiegmann: Influence of an undoped AlGaAs spacer on mobility ehnancement in GaAs-AlGaAs superlattices, Appl.Phys.Lett. 38 (1981) 9, 691-693

|68| Hiyamizu,S., and T.Mimura: High mobility electrons in selectively doped GaAs/n-AlGaAs heterostructures grown by MBE and their applications to high-speed devices, J.Crystal Growth 56 (1982) 2,455-463

|69| Döhler,G.H.: Semiconductor superlattices- a new material for research and applications, Physica Scripta 24 (1981) 2,430-439

|70| Künzel,H., A.Fischer, J.Knecht, and K.Ploog: A new semiconductor superlattice with tunable electronic properties and simultaneously with mobility enhancement of electrons and holes, Appl.Phys. A30 (1983) 1,73-81.

Chapter 2. TECHNOLOGY OF SUPERLATTICE STRUCTURES

Superlattices are multilayer structures fabricated with alternate thin layers (< 20 nm) of at least two materials (or of one material of differently doped regions) grown sequentially upon a suitable substrate. Although superlattice structures can in principle be fabricated for a wide range of material systems, thus offering considerable flexibility, the success in achieving desired properties depends critically upon the control of atomic structure, defects, unintentional and intentional contaminants, etc. Therefore, the critical ingredient for the realization of the potential offered by superlattice structures is their successful growth |1|.

To achieve a desired atomic arrangement (and thus electronic properties of the superlattice) attention has to be focused on growth conditions and growth mechanisms. Unfortunately, very little systematic effort has so far been expended on this aspect of superlattice structures. Such studies would have to be undertaken if the subject is to move beyond solely empirical regime.

A variety of different growth techniques has been used to grow superlattice structures, but so far only few of them seem to be well adapted for satisfying the stringent requirements of such multilayer structures.

Molecular beam epitaxy (MBE) in its essential form |2-4| combined with various innovative concepts of crystallization under ultrahigh vacuum |5-9| offers great hope of achieving nearly perfect structures, desired for research and for device applications. At the base of this technique is the ability to achieve essentially kinetic growth |10-11| as opposed to the near thermodynamic equilibrium growth in more conventional techniques. This kinetic aspect of growth offers the possibility of creating random homogeneous metastable alloys between material systems which may exhibit miscibility gaps under thermodynamic equilibrium conditions |1|. Such alloys can equally well be incorporated into the superlattice structures if desired.

An empirical relation between atomic size mismatch and the occurrence of miscibility gap between binary tetrahedral compound semiconductors has been found for near thermodynamic equilibrium growth. In terms of the lattice constant mismatch of such compounds this translates into a value of approximately 7.5%. An interesting example here is the $InAs_{1-x}Sb_x$ system which exhibits a miscibility gap for $0.3 < x < 0.9$. Making use of the kinetic aspect of the MBE growth one can, however, circumvent the thermodynamic restrictions and fabricate random alloys of this kind over the entire composition range |1|.

The inherent features of the MBE technology make it well suited to the research environment and to the demonstration of new concepts, employing thin layers and abrupt interfaces. However, recently the metalorganic chemical vapour deposition (MO CVD) technique has been increasingly employed by a number of laboratories to fabricate sophisticated multilayer structures, including superlattice structures as well |12-18|.

Many other names have been used for this growth process, including "metalorganic, or organometallic, vapour phase epitaxy" (MO VPE or OM VPE), and "metal alkyl vapour phase epitaxy" (MA VPE) |12|. The term "metalorganic chemical vapour deposition" will be used here since it is a more general description of the process, encompassing the use of any of the metalorganic compounds and the growth of nonepitaxial (polycrystalline or amorphous) films.

MO CVD is a growth process in which the constituent elements for the growth are introduced into the growth reactor in the form of gaseous metalorganic alkyl compounds [e.g., trimethylgallium (TMGa) and trimethylarsine (TMAs), or triethylgallium (TEGa) and triethylphospine (TEP)], and in which the layer growth results from the thermal decomposition (pyrolysis) of these gaseous compounds followed by a chemical reaction between the component elements on the heated substrate platelet |19|.

The physical processes that control the epitaxial growth of materials by MO CVD are not as well understood as in MBE. The detailed chemical reaction paths are not well known and the physical kinetics of deposition have not been studied, so far, in great detail. It seems likely that few differences occur here for the two growth techniques, resulting mainly from the presence of high pressures of carrier gas (H_2 or N_2) in the MO CVD process.

Most of the III-V, II-VI and IV-VI semiconductors can be grown by MO CVD. A persistent problem that occurs in the growth of various materials is, however, the formation of 1:1 adducts in the gas phase owing to addition reactions between certain alkyls and hydrides |20|.

In order to grow superlattice structures also other materials technologies have been used. Among them liquid phase epitaxy and hot wall epitaxy are most popular.

Liquid phase epitaxy (LPE) is a growth process in which the single crystal layer crystallizes from a supersaturated solution |21-25|. The layer growth is accomplished by cooling the liquid solution below the saturation temperature while it is in contact with the substrate. The whole LPE process is comprised of:

(i) exposing a substrate wafer to a solution whose composition is that of the liquidus of the crystallized semiconductor compound,

(ii) cooling the saturated liquid in a strickly controlled manner in order to grow the desired film, and finally

(iii) removing the liquid contacting the grown film in order to terminate the growth process.

When growing a superlattice structure, the LPE process consists of a series of growth steps from different solutions with compositions chosen to suit the growth of the constituent layers of the superlattice. The growth steps are connected by a series of transfer steps of the substrate between the liquids used for growing the superlattice.

Hot wall epitaxy (HWE) is a vacuum film deposition technique characterized by a growth process proceeding under conditions very close to thermodynamic equilibrium. The main feature of HWE is the use of a heated liner (hot wall) which serves to enclose and direct the vapour from the source to the substrate |26|. In this manner: the loss of evaporating material is avoided, the high vapour pressures of the compound or its different components can be maintained, and the difference between the substrate and source temperatures can be reduced to a minimum |27-29|.

Most of the superlattices that have been studied so far have been grown by the MBE technique. This is at present the most useful growth technique for artificial semiconductor superlattice structures. Only very recently, the MO CVD technique, especially when performed at reduced gas pressure |30|, became competitive to MBE.

It seems, therefore appropriate to organize this technological chapter as follows. In section 2.1 the main physical and practical aspects of the MBE technique will be presented. In section 2.2 will be reviewed the present day status of the MO CVD growth technique as applied to the growth of superlattices. Section 2.3 concluding this chapter will present a concise survey of growth peculiarities related to different types of superlattices.

2.1. Molecular beam epitaxy

MBE is an epitaxial growth process by which thin layers of different compounds |2-4|, |31-33| crystallize via reactions between thermal-energy molecular or atomic beams of the constituent elements and a substrate surface which is maintained at an elevated temperature in ultrahigh vacuum. Characteristics of MBE are distinguished in many ways from those of other growth techniques used for preparing superlattice structures:

(i) MBE growth has a low rate of typically 1 μm/h,

(ii) a relatively low growth temperature,

(iii) the ability of abrupt cessation or initiation of growth, as mechanical shutters for each effusion source of the constituent elements can be used,

(iv) the possibility of introducing various vapour species to modify the layer composition and to control the dopant concentration by adding, in an convenient way, different beam sources generating required MB fluxes,

(v) a smoothing of the surface of the growing crystal on an atomic scale,

(vi) the facility for <u>in situ</u> analysis.

As a result of these favorable features, MBE provides epilayers and superlattice structures made of different compounds at a device-quality level.

The basic process of MBE growth is schematically illustrated in Fig. 2.1. The molecular beams are generated under ultrahigh vacuum conditions from Knudsen-type effusion cells |34-37|, whose temperatures are accurately regulated and usually controlled by a computer system |38|, |39|. Intensities of the projected beams are

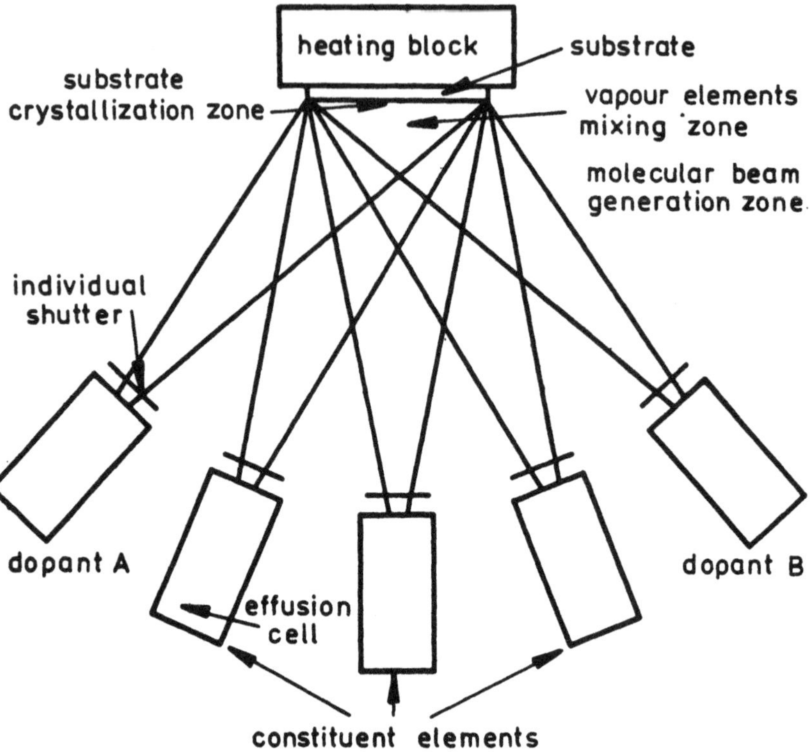

Fig. 2.1. Schematic illustration of the essential elements of a MBE system for fabrication of doped ternary compounds. The three zones where the main physical phenomena of MBE occur are indicated. The illustration borders represent the confines of the ultrahigh vacuum system, (after Ref. 34).

determined by the temperatures of the effusion cells. Choosing appropriately the substrate and cell temperatures, epitaxial films with a desired chemical composition can be obtained. Operation of the shutters permits a rapid changing of the beam species for abrupt alteration of the composition and doping. The uniformity in the film composition and crystal structure depends on the uniformities of the molecular beams across the substrate |37|, |40|,|41|

As it can be seen in Fig. 2.1 the MBE arrangement can be divided in to three zones, where different physical phenomena take place. The first is the MB generation zone. Next is the zone where beams from different effusion sources intersect each other, where the vaporized elements mix, mutually creating a very special gas phase contacting the substrate wafer, on which the crystallization processes take place. The substrate surface, where a lot of kinetic reactions between the constituent elements of the crystallized layer occur can be regarded as the third zone of the MBE physical system.

Let us consider first the physical processes taking place in the first zone |34|.

2.1.1. The total flow rate from effusion cells

The effusion cells used in the MBE systems are usually either cylindrical crucibles of internal diameter of about 1-2 cm and the length of about 5-10 cm with a single circular collimating orifice of diameter d_0 of about 5-8 mm and thickness L_0 of about 0.5 mm in the crucible lid |31|, |34|, or single-channel-type conical crucibles of similar dimensions |37|, |42-44|. For crucible materials both high-purity graphite and pyrolitic boron nitride (PBN) have been used. The advantages of graphite are its low cost and its ready machinability to a design which allows good thermal stability and accurate temperature control. Unfortunately, graphite is more difficult to outgas and chemically more reactive than PBN (for example a graphite cell cannot be used for the evaporation of aluminium). Therefore, PBN is the material chosen for special designs, despite its high cost and the disadvantage that cells made of PBN cannot be finished to a final shape and must be used in an as-deposited form, what may result in less accurate temperature control.

The crucible loaded with the suitable charge, which has to be effused, is placed axially in an alumina resistance heated furnace. In standard effusion cell constructions |31| the helical tantalum heater wire is supported by two concentric Al_2O_3 cylinders. The specially shaped inner alumina cylinder contains the crucible, which provides in its lower part a thermally isolated black body enclosing a butt-welded thermocouple. The thermocouple wires pass through low mass alumina tubes so that the welded end of the thermocouple may touch the bottom of the crucible. The outer alumina cylinder is surrounded by several turns of thin, crimped tantalum foil for an effective thermal isolation of individual cells. Each effusion cell is mounted on an all-molybdenum support with provisions for accurate cell adjustment and rapid dismounting. Tantalum and molybdenum are used exclusively for all of the supports and electrical connections of the cells.

Fig. 2.2 shows an effusion cell used in VG Semicon MBE systems. An important feature of this design is the application of high efficiency, self supporting tantalum radiant heater elements. Together these elements make a large area radiation heater assemblies, capable of high temperature operation. The low thermal mass results in quick response to cell temperature adjustments. A spring loaded W/Re thermocouple provides temperature monitoring and control. The whole cell has an integral water cooled surround, effectively thermally isolating it from the surroundings.

The first theoretical characterization of the distribution of the flux in a molecular beam was performed by Knudsen |45|. He studied effusion from a pinpoint orifice in an effusion cell containing an ideal gas. He showed that the flux incident upon a surface intercepting the molecular beam is proportional to, among other things, the following three terms:

|i| the cosine of the angle between the line normal to the cell's orifice and the line connecting the orifice and the surface,

|ii| the cosine of the angle between the line normal to the surface and the line connecting the surface and the orifice, and

|iii| the reciprocal of the square of the distance from the orifice to the surface.

These results are strictly valid only for true Knudsen cells. An ideal Knudsen cell contains condensed phase and vapour at equilibrium and has an effusion aperture being a small orifice with area a located in an infinitesimally thin cell lid (there is no orifice wall). On the one side of the orifice there is a gas space with a pressure p_1 in it, while on the other side the pressure is p_2 such that $p_1 >> p_2$. The linear dimensions of a are not larger than the mean free path of the gas molecules at either pressure and may be much smaller. The gas reservoirs outside (p_2) and inside (p_1) the cell are so large that the molecules encounter each other more frequently than they do the walls.

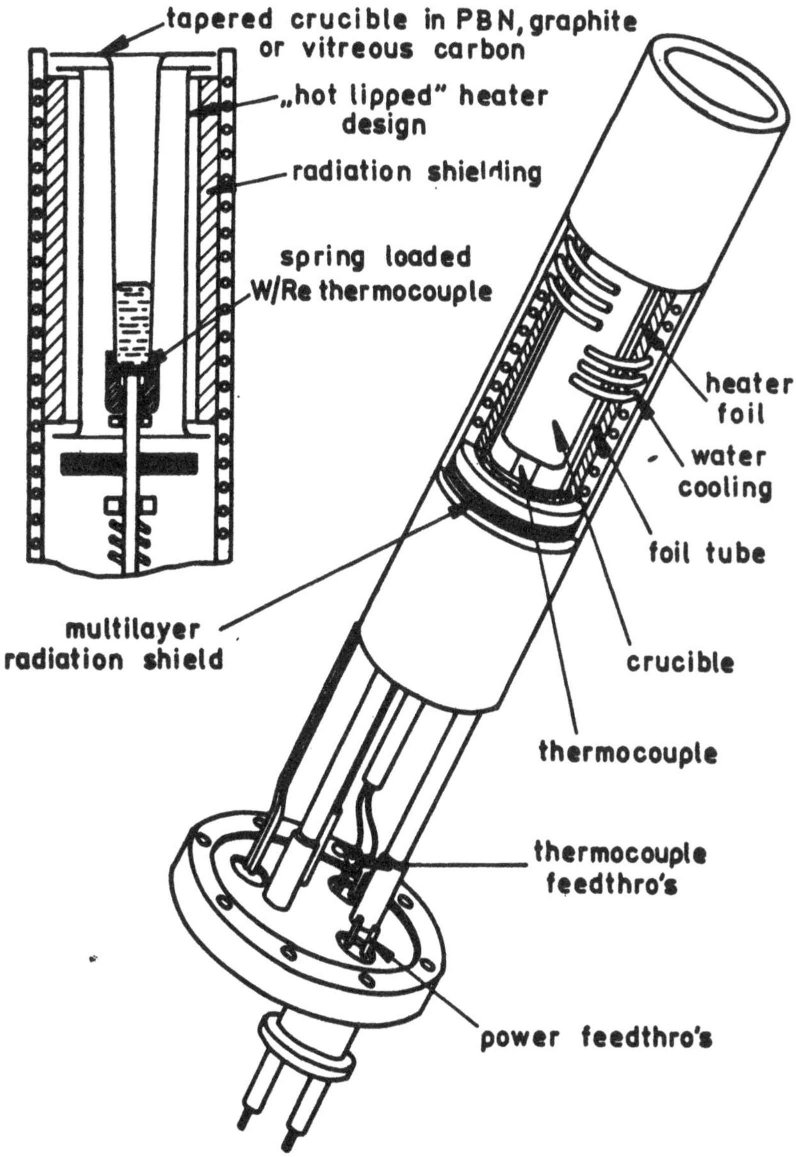

Fig. 2.2. Schematic illustration of a present-day effusion cell used in MBE systems. Section of the central part of the cell assembly (top, left side), and a cut-away diagram of a whole effusion cell assembly (bottom, right side) (after an advertising pamphlet of the VG Semicon company, Great Britain).

Under these conditions, and under the following assumptions:
- molecules are identical point masses,
- interacting forces between molecules are not present,
- velocity distribution of molecules is Maxwellian,
- isotropy of the gas exists,

the net molecular transfer N in unit time taking place through the orifice from the cell to its surrounding is equal to

$$N = \left(\frac{p_1 - p_2}{m}\right) a \left[\mu / (2\pi RT)\right]^{1/2} . \qquad (2.1.1)$$

m is here the mass of a molecule, μ the molecular weight, T absolute temperature (assumed the same on both sides of a) and R the gas constant. If $p_2 = 0$ (as would be the case for effusion into a good vacuum) and p_1, the pressure in the cell is set equal to p, one obtains the well-known Knudsen effusion equation

$$N = (paN_A)/(2\pi\mu RT)^{1/2} , \qquad (2.1.2)$$

where N_A is the Avogadro's number. If p is expressed in torrs, the effusion rate is then |3|

$$N = 3.51 \cdot 10^{22} \; pa/(\mu T)^{1/2} \quad \text{molecules} \cdot s^{-1} \qquad (2.1.3)$$

The results obtained by Knudsen are also useful in treating effusion cells with orifices which have a large diameter. The procedure is to treat the orifice as a combination of many smaller point sources and superimpose their fluxes |45|. Complications arise, however, when the orifice has a significant thickness. In this case the orifice has a somewhat collimating effect on the molecular beam.

Clausing studied this effect for cylindrical orifices |46|, |47|. According to his theory, the effusion equation (2.1.2) should be multiplied by a correction factor W_a called the orifice transmission factor. This factor has the meaning of the probability that a molecule, which enters the orifice in the effusion cell lid at the side of the cell interior, goes to the vacuum resorvoir (in which $p_2 \approx 0$) without having been back-scattered into the cell interior. The Clausing's factor W_a depends only on the geometry of the orifice. As an example, in Table 2.1 some values are given of W_a for a straight tubular orifice of length L_o and diameter d_o

Table 2.1. Values of the Clausing's transmission factor W_a for straight tubular orifices with various orifice dimensions |34|,|47|

L_o/d_o	W_a	L_o/d_o	W_a
0	1.0000	0.75	0.5810
0.1	0.9092	1.00	0.5136
0.2	0.8341	1.50	0.4205
0.4	0.7177	2.00	0.3589
0.5	0.6720	4.00	0.2316

Near ideal Knudsen-cells (equilibrium cells) are seldom used in practical MBE systems (mainly for effusion studies or as dopant cells). In common practice, in order to obtain large epitaxial areas at moderate growth rates, non-equilibrium effusion cells (see, Fig. 2.2) with much larger apertures than typical for the equilibrium cells are required. In this case, the formulae (2.1.2) or (2.1.3) should be multiplied by another correction factor F_c. This factor, however, cannot be calculated directly |31|. Therefore, the emergent molecules must be monitored intermittently using a mass spectrometer, and some kind of feedback system must be employed to regulate the cell temperature.

2.1.2. The role of the source-substrate system geometry in the MBE growth

As it has already been said above, the uniformity in thickness as well as in the composition of the films grown by MBE depends on the uniformities of the molecular beams across the substrate. These quantities depend not only on the individual beam flux at the substrate central point but also on the geometrical relationship between the sources and the substrate |34|.

Let us first consider the problem of the molecular flux distribution across a substrate mounted in the MBE growth chamber axially (Fig.2.3a), and non-axially (Fig.2.3b) against the effusion cell orifice. The number of molecules impinging on the substrate surface per cm^2 and per second at a definite surface point, can be calculated in the following way.

Fig. 2.4 shows the upper part of a near-ideal Knudsen effusion cell. The molecules of the flux generated in this cell escape from

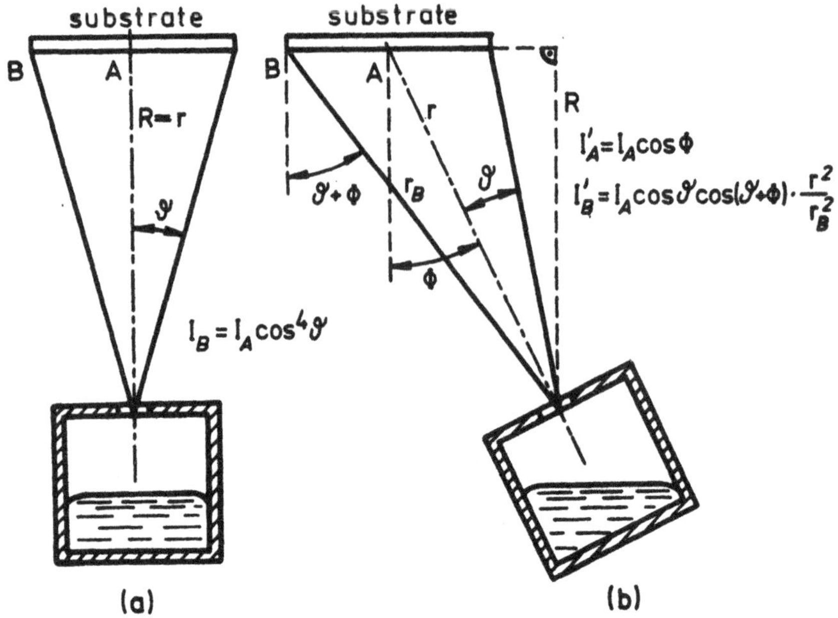

Fig. 2.3 Molecular flux distribution across a substrate mounted in the MBE chamber axially (a) and non-axially (b) against the effusion cell orifice (after Ref. 34).

the cell orifice into the ultrahigh vacuum environment in all directions encompassed by a hemi-sphere with a centre overlapping the orifice centre. The individual molecule m moving along a straight line in the direction of θ toward the cell orifice centre, passes it and escapes into the vacuum in the same direction, because the thickness of the orifice wall is assumed to be infinitesimally thin ($L_o \approx 0$). The total number N_θ of molecules escaping from the orifice area a in the direction of θ into the vacuum is proportional to:

(i) the surface of the orifice seen by the molecule

$$N \sim a \cos\theta$$

(ii) the number of the molecules entering the unit of the orifice area in unit of time

$$N \sim N/a = 3.51\cdot 10^{22} P/(\mu T)^{1/2}$$

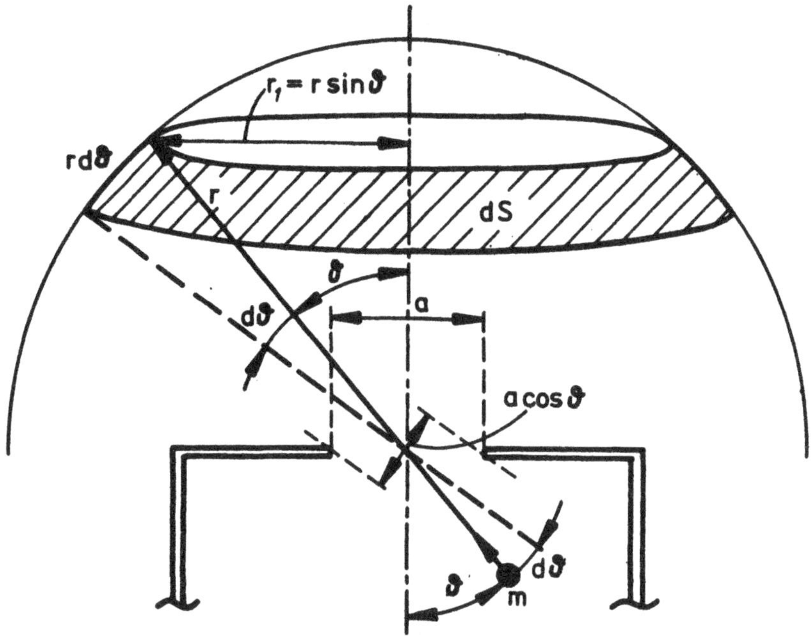

Fig. 2.4 Schematic illustration of the upper part of a near-ideal Knudsen effusion cell, indicating the geometrical quantities relevant to the flux distribution calculations

(iii) the probability that a molecule m enters the orifice in a direction contained between θ and $\theta + d\theta$. This probability is proportional to the solid angle $d\omega$ at dS subtended by these directions. The element $d\omega$ is measured by the area on a sphere of radius r (see, Fig.2.4) of the circular zone between the angle θ and $d\theta$, and so is equal to

$$d\omega = 2\pi \sin\theta d\theta \qquad (2.1.4)$$

The total solid angle into which the molecules escape from the cell orifice is equal to 2π, so the chance that θ lies between θ and $\theta + d\theta$ is

$$P = d\omega/2\pi = \sin\theta d\theta ,$$

thus

$$N_\theta \sim \sin\theta d\theta$$

The number N_θ of molecules escaping from the cell orifice may be expressed by the formula

$$N_\theta = C_o a \cos\theta (N/a) \sin\theta d\theta \qquad (2.1.5)$$

where C_o is a proportionality constant. This constant can be estimated, taking into consideration the obvious fact that θ has to be smaller than, or equal to π/2, and that the total number of molecules escaping from the orifice must be equal to N [given by eq. (2.1.3)]. Thus

$$N = \int_0^{\pi/2} C_o N \cos\theta \sin\theta d\theta = C_o N/2 \qquad (2.1.6)$$

what means that the proportionality constant $C_o = 2$.

The formula (2.1.5) may now be written in the form

$$N_\theta = (N \cos\theta)(d\omega/\pi) \qquad (2.1.7)$$

Using the last formula and the Eq. (2.1.3) the flux of the molecular beam at the central point A of the substrate axially mounted against the effusion cell (see, Fig. 2.3.a) can be easily calculated. It is given by |3|

$$I_A = \left(\frac{N_{\theta=0}}{dS}\right) = \left(\frac{N}{dS}\right)\left(\frac{dS}{\pi r^2}\right) = 1.118\cdot 10^{22}(pa/r^2)(\mu T)^{-1/2}$$

$$\text{molecules}\cdot \text{cm}^{-2}\cdot \text{s}^{-1} \qquad (2.1.8)$$

The flux at the edge point B is equal to

$$I_B = \left(\frac{N \cos\theta}{dS}\right)\left(\frac{dS \cos\theta}{\pi r_B^2}\right)\left(\frac{r^2}{r^2}\right) = I_A(\cos\theta)^4 \qquad (2.1.9)$$

Proceeding in an analogous way the formulae given in Fig. 2.3.b can be derived.

The homogeneities in thickness and in composition of the MBE grown thin film can be expressed quantitatively by appropriate homogeneity coefficients |34|. In order to define these coefficients the geometrical situation shown in Fig. 2.5 should be considered.

The molecular flux at the substrate surface point P(ρ,β) is equal to

$$I_P = I_A \cos\theta \cos(\theta+\phi) r^2 \cos\delta r_P^{-2} \quad \text{molecules cm}^{-2}\text{s}^{-1} \qquad (2.1.10)$$

where δ is the condensation angle. The distance r_P and the angles

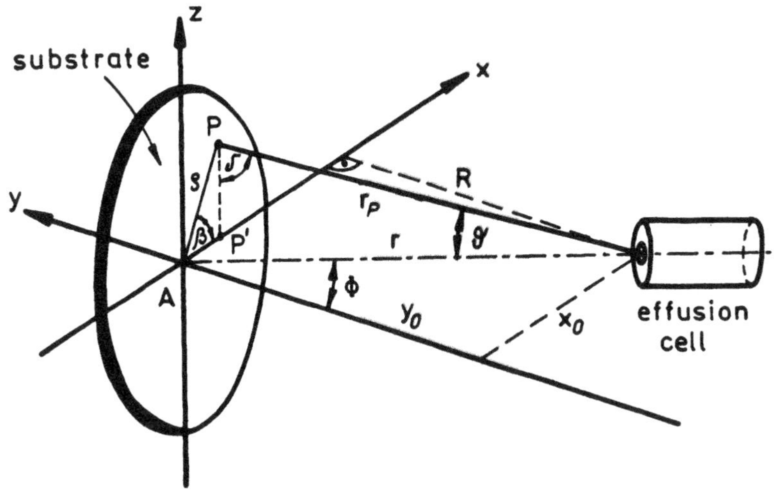

Fig. 2.5. Schematic illustration of the "substrate-effusion cell system" geometry, indicating the polar co-ordinates ρ, β of an arbitrary substrate surface point P (after Ref. 34).

θ and δ depend on the definite values of the two polar co-ordinates ρ and β of the point P on the substrate surface. Consequently, I_P is a function of ρ and β at a given "source-substrate system" geometry. The mass of the species i deposited on the substrate per second and per cm^2 is given by

$$M_i = m_i I_P(\rho,\beta) = \left(\frac{\mu_i}{N_A}\right) I_P(\rho,\beta) \qquad g\ cm^{-2}\ s^{-1} \qquad (2.1.11)$$

where m_i is the mass of one molecule of the species i and the meaning of the other symbols is the same as before. Writing the last expression we have assumed, for simplicity, that each molecule arriving at the substrate surface is incorporated into the epilayer (the sticking coefficient α_{S_i} for the molecules of the species i is equal to 1).

The growth rate of the epilayer consisting of only the species i can be calculated from the formula

$$V_{g_i}(\rho,\beta) = d^{-1} M_i(\rho,\beta) \qquad cm\ s^{-1} \qquad (2.1.12)$$

where d is the density of this species material. Assuming that the beam flux is not dependent on time, the thickness, in cm, of the layer grown from the molecules of species i is equal to

$$l_i(\rho,\beta) = t\, V_{g_i}(\rho,\beta) \qquad (2.1.13)$$

where t is the growth time in seconds.

In the case when molecular beams of the same species i are generated by more than by one source, the resulting thickness of the epilayer in the point $P(\rho,\beta)$ is described by the sum

$$L_i(\rho,\beta,\phi_j) = \sum_{j=1}^{M} l_j(\rho,\beta) \qquad (2.1.14)$$

where j denotes a definite source of the species i. The value of L_i is dependent not only on ρ and β but also on the angle ϕ_j between the axis of the j effusing source and the normal to the substrate surface.

The inhomogeneity in the layer thickness can now be estimated by calculating the thickness homogeneity coefficient

$$\gamma = \frac{L_{i_{max}} - L_{i_{min}}}{L_{i_{max}}} 100\% \qquad (2.1.15)$$

In the case of m different species, each effused from one source (i=j), a composition homogeneity coefficient η can be defined. If the concentration Q_i of the species i in the substrate surface at point $P(\rho,\beta)$ is equal to

$$Q_i(\rho,\beta) = \frac{N_i(\rho,\beta)}{\sum_{j=1}^{m} N_j(\rho,\beta)} \qquad (2.1.16)$$

then the composition homogeneity is described by

$$\eta = \frac{Q_{i_{max}} - Q_{i_{min}}}{Q_{i_{max}}} 100\% \qquad (2.1.17)$$

Estimating now the values of γ and η for different values of X_o,R, ρ and β one can choose the proper geometry of the "substrate-source system". This is, however, very often an extremely difficult task |37|,|40|,|41|. Therefore, if it is only possible, the substrate during the growth is rotated with a constant angular velocity around the axis perpendicular to its surface. The substrate rotation causes a considerable enhancement in thickness and composition layer homogeneities but, unfortunately, excludes the possibility of growth of patterned structures required in many device applications.

2.1.3. Growth processes and doping

Epitaxial growth of semiconductor compounds by MBE involves a series of events, among which the following are most important:

(i) adsorption of the constituent atoms and molecules,

(ii) surface migration and dissociation of the adsorbed particles,

(iii) incorporation of the constituent atoms to the substrate resulting in nucleation and growth.

The thin film grown in this manner has a crystallographic structure related to that of the substrate.

The formation of a film on a solid surface is invariably connected with a phase transition which is characteristic for the chosen growth mode. In the MBE process this is the transition from a vapour phase being far from equilibrium (the partial pressures of constituent beams are, for example in III-V semiconductor compounds systems, for the typical MBE growth conditions, 10 times the equilibrium vapour pressure |11|) to a thin solid film phase.

In this phase transition the above mentioned surface processes should be taken into consideration. These processes are schematically illustrated in Fig. 2.6. The crystal surfaces are divided there into the so called "crystal sites" with which the impinging molecules or atoms may interact. Each crystal site is a small part of the crystal surface characterized by its individual chemical activity. A site may be created by a dangling bond, by a vacancy by a step edge etc. |48|.

The listed surface processes are characterized by a set of relevant kinetic parameters which describe then quantitatively.

The arrival rate is described by the flux (the beam intensity) of the arriving species, which gives the number of atoms impinging on the unit area of the surface per second. For the MBE growth of compound semiconductors, the required fluxes are typically between

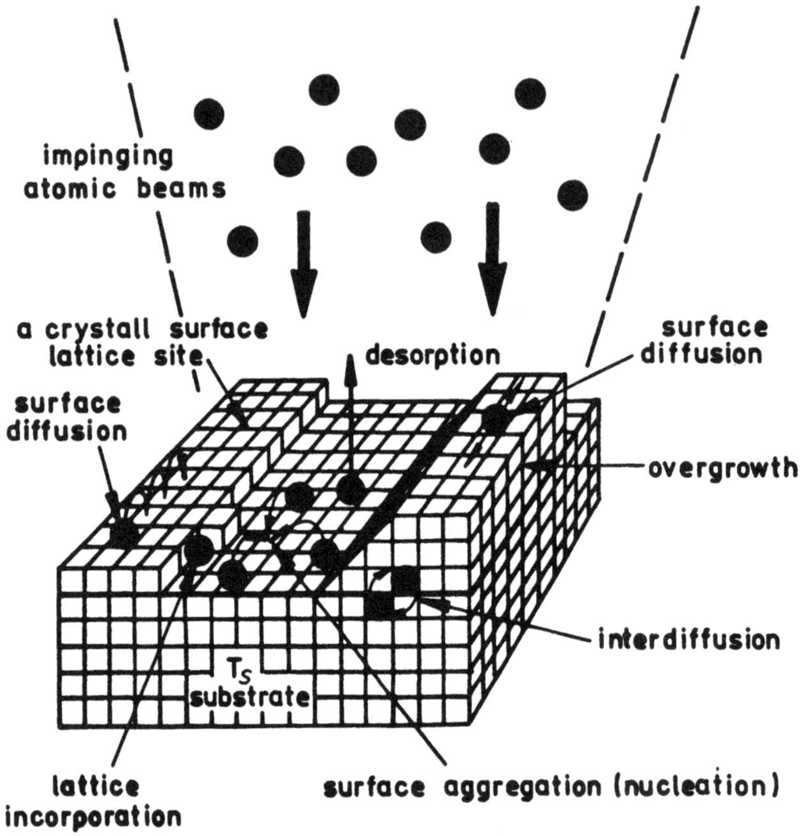

Fig. 2.6. Schematic illustration of the surface processes occurring in growth of thin films by MBE

10^{14} and 10^{16} atoms $cm^{-2}s^{-1}$ |31|. The flux of arriving species can be calculated with Eq. (2.1.3). If the substrate is positioned at a distance r from the effusion cell aperture and is directly in line with it, the number of molecules per second striking the substrate surface of the unit area is given by Eqs. (2.1.8) and (2.1.9).

The atoms arriving at the substrate surface will generally have an energy distribution appropriate to the temperature of their place of origin. In the considered MBE case this will be the ef-

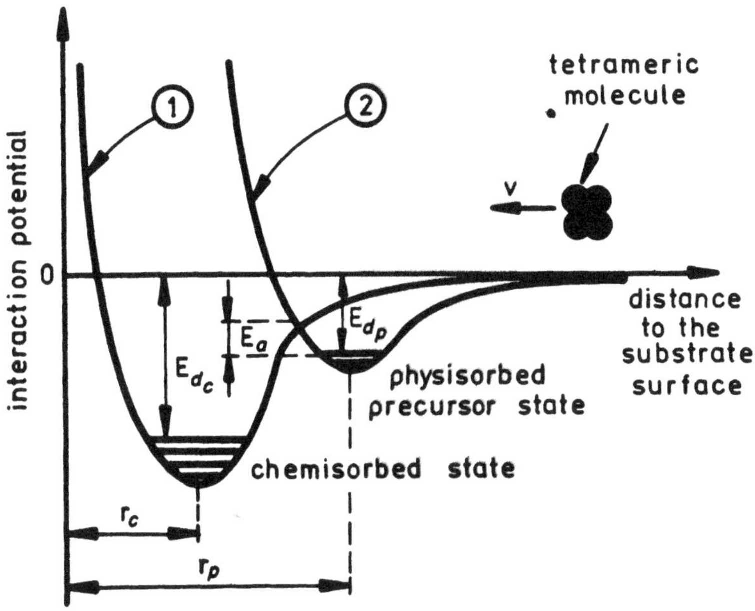

Fig. 2.7. The interaction potential due to the surface, as seen by a molecule impinging perpendicularly to the surface for chemisorbed (curve 1) and physisorbed precursor (curve 2) states.

fusion cell temperature T_i. After arrival at the substrate surface, which has a temperature T_s, usually lower than T_i, the atom may re-evaporate immediately, or may exchange energy with the atoms of the substrate until it reaches equilibrium temperature T_s. The quantitative description of this process is possible by defining the thermal accommodation coefficient, as

$$\alpha_A = \frac{T_i - T_e}{T_i - T_s} \qquad (2.1.18)$$

Clearly, when T_e - the temperature of atoms re-evaporating from the substrate, is equal to T_s, the accommodation coefficient is unity. Thus, it emerges as a measure of the extent to which the arriving atoms reach thermal equilibrium with the substrate |48|.

It is important to differentiate between the accommodation coefficient defined above and the sticking, or condensation coefficient

α_S. The latter is defined as the ratio of the number of atoms adhering to the substrate surface to the number of atoms arriving there

$$\alpha_S = \dot{N}_{ad}/\dot{N}_{tot} \tag{2.1.19}$$

In many cases the sticking coefficient is less than unity and it may be even a very small fraction in cases when the adsorption energy of atoms on the substrate is low, or the substrate temperature is high. Assuming that $\alpha_A = 1$, all the arriving atoms are accommodated on the substrate surface. This does not mean, however, that they will remain there permanently. They still have a finite probability related to the substrate temperature of acquiring sufficient energy to overcome the attractive forces and leave the substrate. If condensation (i.e. aggregation of adatoms) does not occur, all adatoms will eventually re-evaporate. Thus, the sticking coefficient can be almost zero even though the accommodation coefficient is unity. However, if $\alpha_A < 1$, α_S cannot, by definition, be unity.

There are two types of adsorption. The first one is the physical adsorption, called often physisorption, which refers to the case when there is no electron transfer between the adsorbate and the adsorbent, and the attractive forces are van der Waals and/or electrostatic. The second is the chemisorption, which refers to the case when electron transfer, i.e. chemical reaction takes place between the adsorbate and the adsorbent. The forces are then of the type occurring in the appropriate chemical bond. In general adsorption energies for physical adsorption are smaller than for chemical adsorption.

It is obvious that for these two kinds of adsorption different sticking coefficients can be defined. According to the definition of the chemisorption process the sticking coefficient α_{Sc} for the chemisorbed phase may be dependent on the crystallographic orientation of the substrate surface, as well as on the nature and spatial distribution of atoms already adsorbed on this surface. The physisorption, on the other hand, shows generally little or no dependence on surface site arrangement. Consequently, the sticking coefficient α_{Sp} for the physisorbed phase may be taken to be independent of the local environment, i.e. on the orientation and coverage of the surface.

It is experimentally well documented |31| that crystal growth in MBE, in many cases in which molecular species are involved, proceeds through a two step condensation process in which the molecular species reach the chemisorbed state via a precursor physisorbed phase |50|.
The simplest version of such a stepwise condensation process may proceed according to the scheme |51|.

$$A_g \underset{k_d}{\overset{s_p}{\rightleftharpoons}} A_p \xrightarrow{k_a g(\delta)} A_c \qquad (2.1.20)$$

where A_g, A_p and A_c represent the adsorbate in the gas, precursor and chemisorbed states, respectively. The rate constants s_p and k_d apply to condensation and desorption of the precursor state, and $k_a g(\delta)$ is the rate constant for conversion from the precursor state into the chemisorbed state. The conversion rate is assumed to be proportional to $g(\delta)$, the probability of having suitable vacant sites in the chemisorbed state, where δ is caunted in monolayers of the surface coverage.

Another possible version of the condensation process via the precursor state may be sketched as shown below |52|

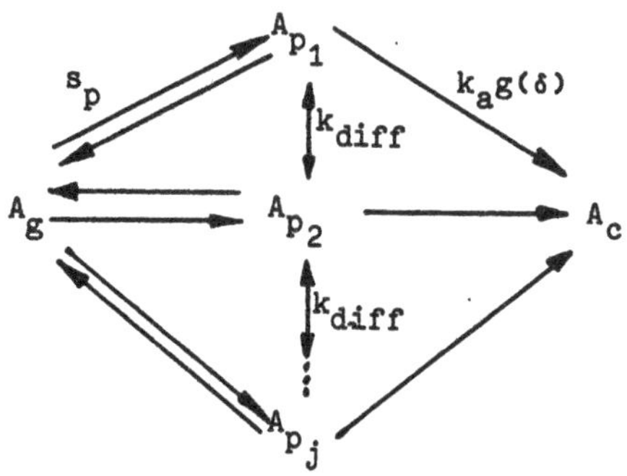

(2.1.21)

This model is similar to that implied by the scheme (2.1.20) except that if the site visited by the molecule in the precursor state is occupied by a chemisorbed species, the molecule is allowed to diffuse over the surface with a rate constant k_{diff} to another site.

The interaction potentials due to the surface seen by an incoming molecule perpendicular to the surface, for chemisorbed and precursor states, are shown schematically in Fig. 2.7. It is evident from that figure that the molecule adsorbed in the precursor state has to overcome a lower barrier when it is subsequently chemisorbed at the surface, than in the case when it is re-evaporating into the vacuum, because $E_a < E_{d_p}$. The illustrated situation is a special example, however, a variety of other potential configurations can also exist.

The above discussion should make it clear that the ability of the atoms to move on the surface, their mean residence time, and the flux of atoms and molecules at the growing solid surface are kinetic parameters which influence the MBE growth process in a decisive way |11|.

In order to gain an understanding of crystal growth with MBE one has to employ the appropriate surface-analytical techniques |53|. Such techniques include reflection high energy electron diffraction (RHEED), Auger electron spectroscopy (AES), secondary ion mass spectroscopy (SIMS), and others. Most research, and many technological MBE systems now contain several components of these surface-analytical tools.

The RHEED technique provides a very sensitive tool for observing surface structure, microstructure and smoothness, for indicating changes in the structure of the outermost layers of atoms of the epitaxial film during growth as a function of growth parameters, and for obserwing surface topography (especially at the early stages of growth).

The AES technique is used for: characterization of the chemical composition of the individual substrate surface, verification of surface accumulation of dopants during epitaxial growth, determination of the chemical composition of the grown layer.

SIMS technique is a powerful tool for the determination of the chemical composition of the outermost atomic layers of a solid. Compared to AES the main advantages of SIMS are: high absolute sensitivity for many elements, good elemental specifity, high mass resolution. When high erosion rates of about 1 nm/s are chosen by applying high primary ion currents, this technique can also be used for the analysis of bulk material and for surface profiling. However, the quadrupole mass spectrometer, appearing in the SIMS system, can be utilized for effective diagnosis of the gas ambient as well as for the investigation of source material composition and of desorption phenomena occurring at the epilayer surface.

Surface kinetic data on growth and doping can be obtained by using the modulated molecular beam (MMB) technique |54-57|. In this technique (see, Fig. 2.8) perturbation of an incident molecular

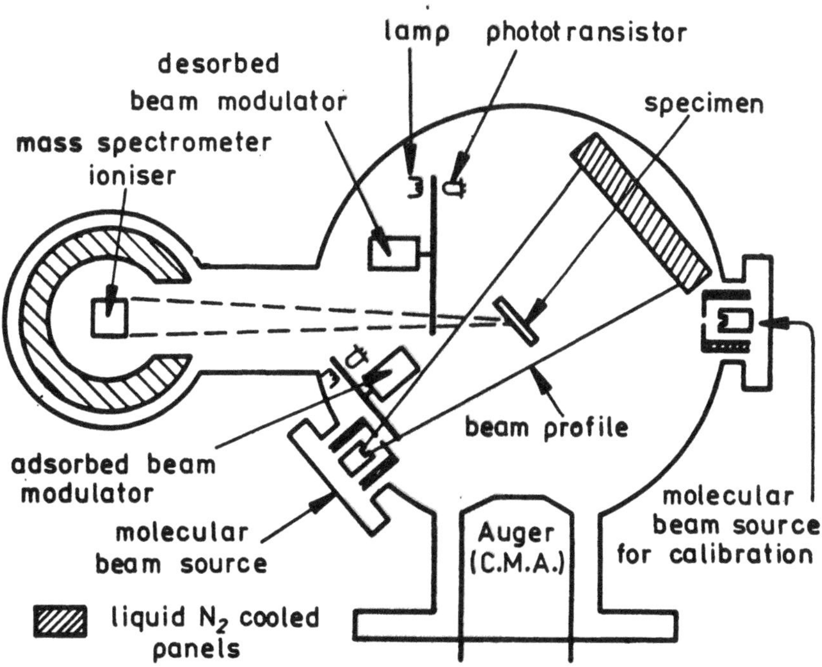

Fig. 2.8. Experimental arrangement for modulated molecular beam studies, as used by Foxon and Joyce (after Ref. 54).

flux gives rise to a time varying concentration of adsorbed molecules on the substrate surface, and hence to a time dependent desorption rate, which can be measured mass spectrometrically. Alternatively, by modulating the desorption flux, atomic and molecular species which are leaving the surface directly can be identified and distinguished from background vapour species. Additional utilization of AES and RHEED techniques permit detailed assessment to be made of substrate surface at any stage of a process, and combined with temperature programmed thermal desorption provide information on surface stoichiometry.

The most closely studied material system in MBE has been, so far, gallium arsenide. Therefore, this material will serve here as an example for illustrating the growth and doping phenomena characteristic for MBE.

There are two procedures which may be used for the preparation of GaAs films by MBE. In the first, the film is grown from beams of tetrameric arsenic molecules (As_4) and of Ga atoms. In the second beams of dimeric arsenic molecules (As_2) and of Ga atoms deliver the constituent atoms of the grown film to the heated substrate. The As_4 molecular beam is generated in an effusion cell charged with pure arsenic, while the As_2 molecular beam is effused from a cell containing GaAs.

The chemistry of surface reactions in both of the growth procedures has been investigated in detail by Foxon and Joyce |54|,|56| with the MMB technique. They formulated kinetic models which are consistent with the experimental data obtained for each of the considered crystallization processes of GaAs.

In the first growth procedure, As_4 molecules are adsorbed into a mobile precursor state and in the absence of a surface Ga population have a sticking coefficient equal to zero, but a measurable surface lifetime. The temperature dependence of this lifetime reveals a desorption energy of the As_4 molecules equal to about 0.4 eV. With a coincident Ga flux, As_4 exhibits a temperature independent sticking coefficient (in the temperature range between 450 and 600 K) which is, however, a function of I_{Ga}, the Ga beam intensity. The desorption of As_4 is a second order process with respect to its adsorption when $I_{Ga} >> I_{As_4}$, and is the prevailing process when $I_{Ga} << I_{As_4}$. The sticking coefficient of As_4 is always less or equal to 0.5, even when $I_{Ga} >> I_{As_4}$. However, when $I_{Ga} << I_{As_4}$ one As atom stick for every Ga atom supplied.

These results can be explained by a process of dissociative chemisorption, with a pairwise interaction of As_4 molecules adsorbed on adjacent Ga lattice sites (see, Fig. 2.9a). When the As_4 surface population is small, compared to the number of Ga sites, the rate limiting step is the encounter/reaction probability between As_4 molecules leading to the second order kinetics. As the As_4 surface

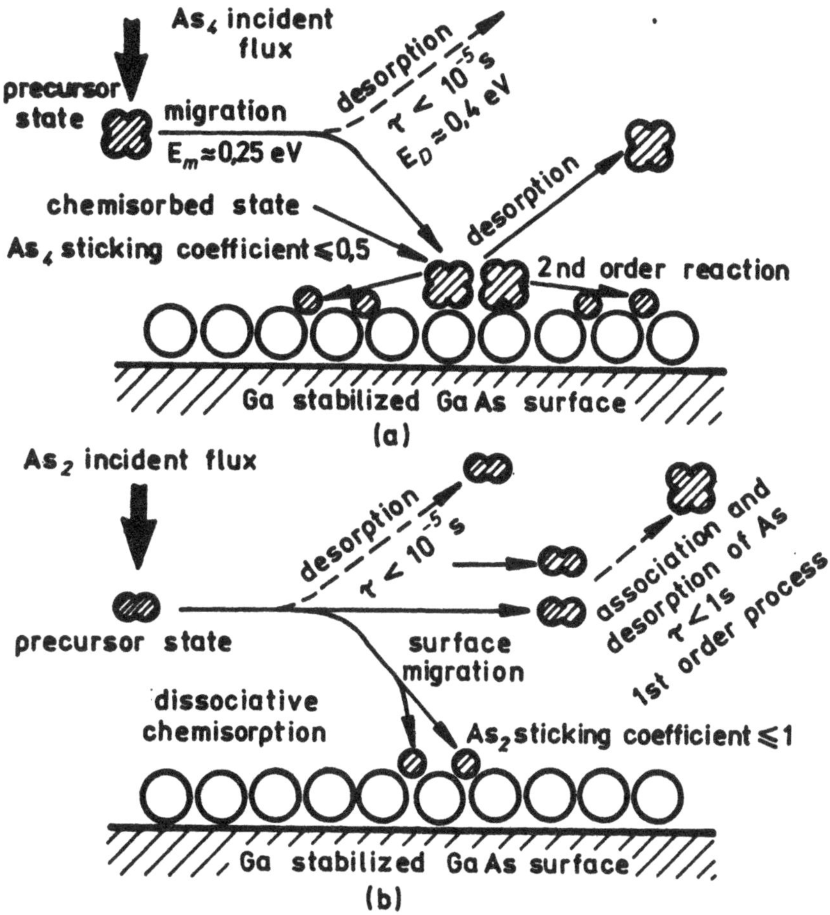

Fig. 2.9. Main features of the growth models for GaAs layers crystallized by MBE on GaAs substrates (after Foxon, Ref.58). a) Growth from As_4, and Ga beams below 600 K, and b) growth from As_2 and Ga beams below 600 K.

population is increased, there is an increasing probability that an arriving molecule will find adjacent sites occupied and the desorption rate becomes proportional to the number of molecules being supplied (a first order process).

In the second growth procedure, As_2 molecules are first adsorbed into a mobile, weakly bound precursor state (see, Fig. 2.9b). Dissociation of adsorbed As_2 molecules can only occur when the molecules encounter paired Ga lattice sites while migrating on the surface. In the absence of free surface Ga adatoms, As_2 molecules have a measurable surface lifetime but no permanent condensation occurs. The sticking coefficient of As_2 molecules is thus a function of the arrival rate of Ga atoms I_{Ga}, being unity only for a monolayer of Ga atoms or for $2I_{Ga} > I_{As2}$. From these kinetic processes it follows that stoichiometric GaAs can be grown for relative fluxes (beam intensities) of As_2 and Ga larger than, or equal to unity.

The presented growth models describe well the real physical situation, when the growth temperature of GaAs is lower than, or equal to 600 K. During the growth of GaAs at growth temperatures above 600 K an additional process occurs on the substrate surface. As_2 is lost by desorption from the GaAs surface at a rate which increases with increasing temperature. This desorption process creates a Ga adatom population on the surface and in the absence of an impinging arsenic flux results in a Ga-stable structure being established. In the limit free gallium may be produced on the substrate surface. As_2 lost in this way from the surface may be replaced by dissociation of incident As_2 or As_4 beams to maintain an As-stabilized structure providing that the arsenic supply rate exceeds the rate of loss of As_2 by thermal desorption. The flux ratio I_{As4} to I_{Ga} required to maintain a specific surface structure during growth will therefore depend upon the substrate temperature and the absolute magnitude of the fluxes. The growth rate of MBE crystallized GaAs films is entirely controlled by the flux density of the Ga beam impinging on the substrate surface. This implies that nearly all the Ga atoms incident on the surface get incorporated into the growing epitaxial layer.
The growth model established from kinetic measurements is not unique to GaAs but is also valid for AlAs, InP, for a number of other III-V compounds, and with minor modifications also for ternary III-III-V alloys |31|.

In practical terms, $Al_xGa_{1-x}As$ films having good crystallographic perfection can be grown by simply directing beams of Ga,Al and excess As_4 simultaneously to the growth interface. Since the sticking coefficient of Al on GaAs is unity |31|, a good compositional control of the growing film can be achieved by adjusting the flux densities of the impinging group III beams.

From a thorough kinetic study of the involved surface processes Foxon and Joyce |59| have deduced that the principal limitation to the growth of III-III-V ternary alloys by MBE is the thermal stability of the less stable of the two III-V compounds of which the alloy may be considered to be composed. Thus, the surface composition of the III-III-V alloy reflects the relative flux ratio of the group III elements only, if growth is carried out at temperatures below which GaAs in the case of $Al_xGa_{1-x}As$, or InAs in the case of $Ga_xIn_{1-x}As$, or InP in the case of $Ga_xIn_{1-x}P$ are thermally stable. This explains why a transfer of the growth parameters previously determined for GaAs to $Al_xGa_{1-x}As$ system works without complications concerning the compositional uniformity of the alloy. AlAs is more stable than GaAs, so GaAs thermal stability determines the compositional uniformity of $Al_xGa_{1-x}As$, grown by MBE. It should also be added, that theoretical estimations based on an atomistic Monte Carlo simulation procedure |60| have shown that surface migration of the group III elements play the most important role in controlling the alloy surface quality in the III-III-V material system grown by MBE.

Uniform GaAs and $Al_xGa_{1-x}As$ layers are now routinely grown by MBE on substrate areas as large as 30 x 30 mm. When grown with an As-stabilized (2 x 4) or c (2 x 8) surface reconstruction |61-63| at substrate temperatures ranging from 450 to 610°C, a smooth mirrorlike surface appearance is generally achieved |31|.

Dopant incorporation in MBE growth is usually achieved by using additional effusion cells which contain the appropriate doping element. The resulting electrical and optical properties of the grown films depend basically on two processes. First, they rely on the flux of the impurity atoms reaching the growth interface, on their sticking probabilities and on their surface lifetimes, particularly in the presence of large fluxes of the constituent elements of the grown film. The second process is the actual incorporation behaviour and the extent of electrical activity of the dopant.

Many of the dopants which are useful in other forms of epitaxy do not behave well with MBE. All materials which have been used successfully as dopants in MBE have unity sticking coefficients (as measured by the MMB technique) over the whole range of growth conditions. In order to be more specific we will again restrict our considerations to GaAs.

At present time four suitable acceptor impurities have been found for GaAs grown by MBE. These are: beryllium, manganese, germanium and ionized zinc |31|. The most ideal p-type dopant is beryllium, for which p-type doping levels up to $2 \cdot 10^{20}$ cm^{-3} in GaAs have been achieved with a perfect surface morphology of the doped epilayer |64|. There are no complications during Be incorporation arising from segregation or anomalous diffusion effects. The values of free acceptor concentrations are directly proportional to the Be effusion cell temperature, i.e., to the Be vapour pressure. Each incident Be atom produces one free acceptor, which provides a fairly shallow acceptor level in GaAs only 30 meV above the valence band-edge.

The incorporation of donor impurities in MBE grown GaAs seems to be less complicated as compared to acceptor incorporation. Six donor impurities for MBE of GaAs have been investigated so far, however, only three elements, i.e., silicon germanium and tin are important for practical applications |31|. The most convenient n-type dopant for GaAs is tin. The Sn doped GaAs films prepared by MBE exhibit a very low degree of compensation. The free donor concentration in GaAs is directly proportional to the Sn flux in the molecular beams, and concentrations up to $5 \cdot 10^{19}$ cm^{-3} can be easily achieved. The only major disadvantage of using Sn is its tendency to accumulate at the GaAs surface during MBE growth |31|, |65|. This is the reason why silicon has recently become the most intensively investigated n-type dopant for MBE grown GaAs and $Al_xGa_{1-x}As$ |66-69|.

Although Si is an amphoteric dopant for LPE grown GaAs, in GaAs grown by MBE it is mainly incorporated on Ga sites yielding n-type material of fairly low compensation. The reaction describing the Si incorporation into GaAs may be written as |10|

$$Si(g) + V_{Ga} \rightarrow Si^{+}_{Ga} + e \qquad (2.1.22)$$

where Si(g) is silicon in the gas phase, V_{Ga} is a vacant gallium site in GaAs, Si^{+}_{Ga} is a silicon positive ion on a normal gallium site in GaAs and e is an electron in the conduction band of GaAs. It is worth-while to mention that for useful dopant fluxes $\geqslant 10^9$

atoms $cm^{-2}s^{-1}$ obtained by sublimation, the Si source must be heated to temperatures ≥ 1000°C. Therefore, the material used for the Si effusion cell must be carefully selected (PBN is used most frequently), since cells operated at such high temperatures are potential sources of unintentional impurities.

Concluding the considerations concerning the MBE growth and doping processes it should be said that MBE can be used to achieve extreme dimensional control in both chemical compositions and doping profiles. Single crystal superlattice structures with layer thickness of only a few atomic layers can be easily grown with many semiconductor materials. The extraordinary versatility of this epitaxy technique, coupled with the recent appearance of multi-chamber, multi-function MBE systems (see, Fig. 2.10) makes it one of the most important device fabrication technique.

2.2. Metalorganic chemical vapour deposition

The term "metalorganic" denotes a broad class of compounds that contain metal-carbon bonds (known as organometallic compounds), as well as those with metal-oxygen-carbon bonds (known as alkoxides) and the coordination compounds of metals and organic molecules |12|. Although many of the chemical and physical properties of a large number of metalorganics were established in the 1950's, it was not until 1968 that any work was reported on the growth of epitaxial semiconductor thin films with metalorganic sources. Manasevit, showed for the first time |70| that vapour phase mixtures of the metalorganic compound trimethylgallium $(CH_3)_3Ga$, and the hydride arsine, AsH_3, when pyrolized in the temperature range from 600 to 700°C in an H_2 atmosphere, could be used to grow thin epitaxial films of GaAs in an open tube reactor (see, Fig. 2.11).

At the same time, research was being carried out on the application of MO CVD to the growth of other homoepitaxial and heteroepitaxial semiconductor thin films. However, because of the availability of different, more advanced technologies for the growth of III-V compound semiconductor thin films, such as LPE and halide-transport VPE process, the MO CVD technology was not extensively used for this purpose. Only in the past five years, there has been an increasing interest throughout the world in MO CVD growth of semiconductor device structures. This is the result of the in-

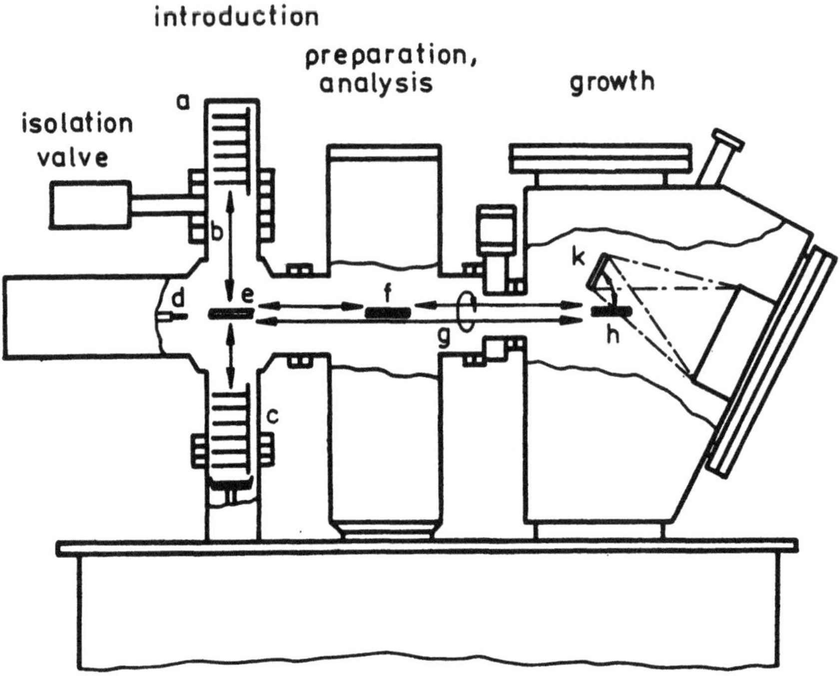

Fig.2.10. Schematic drawing of the Perkin-Elmer MBE system PHI model 425B (after an advertising pamphlet of the Perkin-Elmer Company, USA). The operation of substrate handling hardware is shown. Cartridge is loaded into the introduction chamber (a) which is then evacuated. The gate valve is opened, the elevator engages the cartridge (b), and lowers it into the storage chamber (c). The elevator is raised, the fork (d) extracts the selected substrate from the cartridge (e), and transports it to the preparation chamber, where it can be transferred to the stage (f), or the growth chamber. When moved to the growth chamber, it is inverted (g), transferred to the growth stage (h), and moved to the growth position (k).

terest in the potential large-scale production of certain compound semiconductor devices based on submicrometer-thick layers (for instance, field effect transistors, and high quality optoelectronic heterojunction devices, like lasers, photodetectors and solar cells). The MO CVD technology is well-suited for the fulfillment of these requirements and has some advantages over alternative

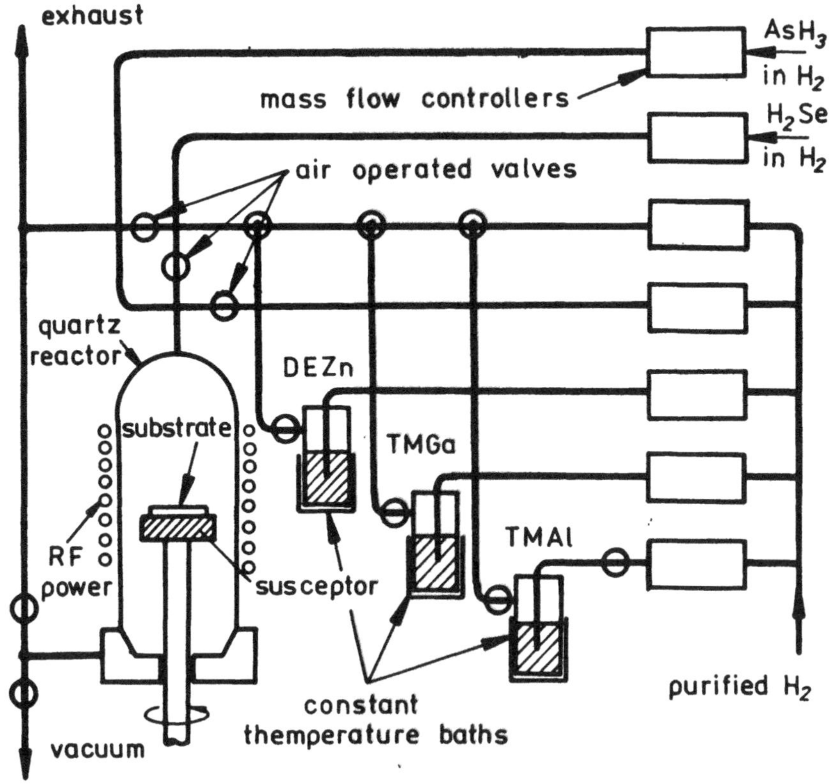

Fig.2.11. Schematic diagram of a vertical atmospheric-pressure MO CVD reactor for growing doped $Al_xGa_{1-x}As$ epilayers (after Dupuis, Ref. 12).

materials technologies |17|,|18|.

The crystallization in the MO CVD process occurs by passing a homogeneous gas-phase mixture of reactans and carrier gases over a heated substrate in a cool-wall reactor. Only the temperature of the substrate needs to be carefully controlled. Most of the properties of films grown by MO CVD are only weakly dependent on temperature, so that slight variations in substrate temperature are inconsequential.

The partial pressure of the various gaseous constituents can be controlled by electronic mass flow control of the source flow rates. These features allow all of the critical variables in the growth process to be controlled with extremely high accuracy, ensuring reproducibility.

The metaloorganics of interest for the growth of semiconductor films are typically liquids at room temperature, although some of the useful metalorganic compounds are solids at room temperature and above. These metalorganics generally have relatively high vapour pressures and can be readily transported into the reaction zone by passing a carrier gas, such as H_2, through the liquid sources or over the solid sources. The hydrides of interest for semiconductor film growth are gases at room temperature and are generally used as dilute mixtures in a H_2 matrix. These metalorganic and hydride compounds are mixed in the vapour phase and are usually pyrolyzed in a flowing H_2 atmosphere in an open tube reactor operating at atmospheric pressure or at a reduced pressure of about 70 torr. Pyrolysis temperatures typically in the range from 600 to 800°C are employed. Energy for heating the gaseous source mixtures is usually provided by a high-power radio-frequency generator operating at about 450 kHz. This RF generator heates a graphite "susceptor" on which the single crystal substrate is placed. The gas mixture at, or near, the substrate surface is thus heated to high temperatures while the walls of the reactor chamber remain relatively cool, leading to the deposition of the semiconductor film on the substrate crystal and to a small loss of reactants to the surfaces of the reactor chamber.

Multilayer, multicomponent epitaxial structures can be deposited in a single growth sequence. The reactors typically used for MO CVD growth have facilities for several metalorganic and hydride sources. In addition, the process involves no etching species, so that abrupt interfaces between various materials can be formed. The absence of an etching species also helps to achieve the growth of films with uniform thickness and composition, since the growth process is not result of competing deposition and etching reactions, as is the case for some other vapour-phase epitaxy processes. The process is scalable to high volume |71|. Large-area uniform growth of epitaxial structures can be achieved by MO CVD with equipment similar to that currently used in large-scale commercial epitaxial growth of silicon. In addition, when desirable, high growth rates compatible with production applications are achievable.

Most of the research on MO CVD growth of compound semiconductors has employed the methyl and ethyl metalorganics. These compounds are relatively simple to prepare and can be readily pyrolized in a hydrogen atmosphere to yield the corresponding metal atoms, with methane or ethane produced as a by-product. Table 2.2 lists the principal metalorganics that have been utilized in the growth of compound semiconductor films. All of these compounds are now commercially available in "electronic grade" purity |12|.

Table 2.2. Metalorganics for semiconductor thin film deposition (after Dupuis, Ref. 12).

Group of metal in periodic table of elements	Compound	Symbol
IIa	Biscyclopentadienylmagnesium	Cp_2Mg
IIb	Dimethylzinc	DMZn
	Diethylzinc	DEZn
	Dimethylcadmium	DMCd
	Dimethylmercury	DMHg
	Diethylmercury	DEHg
IIIa	Trimethylaluminum	TMAl
	Trimethylgallium	TMGa
	Triethylgallium	TEGa
	Diethylgallium chloride	DEGaCl
	Trimethylindium	TMIn
	Triethylindium	TEIn
IVa	Tetramethyltin	TMSn
	Tetraethyltin	TESn
	Tetramethyllead	TMPb
	Tetraethyllead	TEPb
Va	Triethylphosphine	TEP
	Trimetylantimony	TMSb
	Trimethylarsine	TMAs
VIa	Dimethyltelluride	DMTe
	Diethyltelluride	DETe

The large variety of metalorganic sources that are listed in Table 2.2 implies that a wide variety of compound semiconductor materials has been grown by MO CVD. In fact, binary compound semiconductors have been grown in the III-V, II-VI and IV-VI semiconductor materials |72|. Also, many of the important ternary and quaternary III-V compound semiconductors have been grown by MO CVD.

2.2.1. The growth mechanism in MO CVD

The net chemical reaction for the specific case of MO CVD growth of GaAs from the TMGa and AsH_3 may be written in the form

$$(CH_3)_3Ga + AsH_3 \xrightarrow[700^oC]{H_2} GaAs + 3CH_4 \qquad (2.2.1)$$

Similar reactions are employed for the growth of other binary, ternary, and quaternary compound semiconductor thin films. For example, for $Al_xGa_{1-x}As$ growth, the following process is typically employed

$$(1-x)[(CH_3)_3Ga] + x[(CH_3)_3Al] + AsH_3 \xrightarrow[700^oC]{H_2} Al_xGa_{1-x}As+3CH_4 \qquad (2.2.2)$$

In this case, the alloy composition of the epitaxial film is directly related to the relative initial partial pressures of the TMGa and TMAl in the vapour phase. Unfortunately, this simple relation between the composition of the sources in the vapour phase and composition of the resulting solid film is not always true for the growth of compound semiconductors by MO CVD.

The important features in the preparation of materials, i.e., compositional control, growth rate, crystal quality, purity and electrical as well as optical properties, are determined by the chemistry of the process involved |72|. The extension of the MO CVD technology to each new material has usually involved new chemical problems, and there is still a need for improved understanding of the basic mechanisms of growth.

Since the early work of Manasevit |70|, the involvement of researchers in attempts to achieve an understanding of the basic processes has been small compared with the intensive effort that has been directed towards the application of the MO CVD process for the growth of device structures. Recently, however, there has been a growing realisation of the potential of MO CVD for a wider range of materials and, as a result of research into the chemistry involved, consideration has been given to organometallic compounds other than simple alkyls as precursors. One particular group of compounds that attracted much interest is the adducts of group III compounds.

An adduct is a compound formed between an electron acceptor molecule (Lewis acid) and an electron donor molecule (Lewis base) |20|. The strenght of the donor bond is less than the covalent metal-carbon bond and is highly dependent on the organic groups attached to the constituent elements of the crystallized compound |73|. In the case of III-V compounds the adducts are created according to the reaction

$$\underset{\text{(Lewis acid)}}{MR_3} + \underset{\text{(Lewis base)}}{XR_3} \rightarrow \underset{\text{(adduct)}}{MR_3 \cdot XR_3} \quad (2.2.3)$$

where R stands for alkyl, M for metal, and X for group V element. This reaction is for many adducts reversible and they dissociate on heating.

The development of adducts as alternative precursors to alkyls in MO CVD of III-V semiconductor compounds has been stimulated by the difficulties that have been encountered in obtaining reproducible layers of In-containing compounds. Furthermore, adducts are more easy and safe to handle than alkyls are, what makes it possible to simplify growth systems and hence to reduce their cost.

Considering the system (In, Ga)(As, P) used for longer wavelength optoelectronic applications one encounters still a good deal of uncertainty about the optimum conditions for growth and good reproducibility and uniformity have still to be achieved. The problems are accentuated both by the need to ensure good lattice-matching in heterostructures, and by the developing requirement for thin-layer and superlattice structures.

Although moving from GaAs to InP appears to involve only a simple replacement of TMGa by TMIn, and AsH_3 by PH_3, the difference in chemical behaviour of the gas phase formed after mixing the TMIn and PH_3 gases caused problems. Homogeneous gas phase reactions which caused no serious problems and were, in fact, not even considered for the growth of Ga and Al containing semiconductor materials, were found to completely dominate the growth process for In containing materials. These reactions may lead, in the case of crystallization of InP, to the formation of the unstable adduct $InR_3 \cdot PH_3$, which spontaneously decomposes forming an involatile polymer $(InRPH)_n$ and gaseous CH_4. The polymer condenses on the inlet to the deposition zone of the reactor resulting in lower growth rates and inhomogeneous composition of the grown epilayer. Moreover, the decomposition of the adduct is catalysed by the walls of the reactor inlet and the extent of the decomposition is sensitive to small changes in the nature of the surface, giving variations from day to day and from system to system. InP grown under these conditions was found to be of minor and of variable quality both in electrical properties and in morphology |20|.

The parasitic gas phase reactions could be avoided by using low pressure reactors or by using adduct forms of the MO source of In |74|. The growth of InP from the $In(CH_3)_3 \cdot P(C_2H_5)_3$ adduct may serve here as an example.

In this case the In alkyl is first reacted with a Lewis base, i.e. with triethylphosphine, to form the stable 1:1 adduct $In(CH_3)_3 \cdot P(C_2H_5)_3$ according to the reaction (2.2.3). The donor bond is here relatively strong because there are no reactive hydrogen atoms bonded to the phosphorus which may cause the facile alkane elimination and subsequent creation of a reactive polymer $(InRPH)_n$. Therefore, addition of PH_3 to this adduct does not result in the formation of the unstable $InR_3 \cdot PH_3$ adduct, but leads to crystallization of InP according to the reaction |20|

$$In(CH_3)_3 \cdot P(C_2H_5)_3 + PH_3 \rightarrow InP + 3CH_4 + P(C_2H_5)_3 \qquad (2.2.4)$$

TEP was found to be particularly suitable for preparing the stable 1:1 adduct because of its high thermal stability. It remains largely undecomposed in a system where the residence time of the species over the heated substrate is short. Thus, if AsH_3 is used in the reaction (2.2.4) in place of PH_3, indium arsenide can be grown without any apprecible incorporation of phosphorus in the InAs epilayer.

2.2.2. Crystallization in the MO CVD process

The MO CVD process for the growth of a semiconducting compound is in general realized in subsequent steps usually taking place in different zones of the reactor. Discussing this process in detail, we will again (as it was in the case of MBE) restrict our considerations to GaAs and $Al_xGa_{1-x}As$ compounds.

In the MO CVD process, a nonequilibrium situation is intentionally created where the input gas phase has a higher concentration of reactants than would exist in equilibrium with the substrate at the growth temperature. Thus, the first consideration in understanding the process is thermodynamics.

At equilibrium, for each component the chemical potentials μ_i must be equal in the vapour and solid phases. When the system is not in equilibrium, the thermodynamic driving force to restore equilibrium is $\Delta\mu$. The maximum quantity of solid which can be produced is simply the amount which would establish equilibrium. Thus, the maximum growth rate is fundamentally limited by thermodynamics and the total rate at which the supersaturated gas is passed through the reactor |75|.

Except at very high temperatures, the growth rate is considerably lower than that calculated from thermodynamics. The kinetics of the individual growth processes are not fast enough to allow equilibrium to be established throughout the system. Usually, equilibrium is established at the growing surface, but not in the gas phase far from the substrate. The growth rate is therefore directly proportional to the difference $\Delta\mu$ in chemical potential between the input gas phase and that part of the gas which is in equilibrium with the solid being grown.

Another important consequence of thermodynamics is often the control of the composition of the grown solid III-V ternary or quaternary alloy. When thermodynamic equilibrium is established at the growing interface, thermodynamics allows the determination of the solid composition which will be grown from the vapour of a given composition at a specified temperature.
If the gas phase is considered to be an ideal gas, the chemical potential of the i-th component in the gaseous mixture is given by

$$\mu_i = \mu_i^o + RT \ln(p_i/p_i^o) \qquad (2.2.5)$$

where R and T have their usual meaning and p_i is the partial pressure of the i-th component. The superscript "o" in this case refers to the pure component in the gas phase.

An accurate description of the termodynamics of the solid phase must deal with its monideality, i.e., for the solid

$$\mu_i = \mu_i^o + RT \ln a_i \qquad (2.2.6)$$

where a_i is the activity. This quantity is often defined as the product of concentration x_i, and the activity coefficient γ_i, which is calculated in terms of the interaction parameter |75|. This parameter may depend on the difference in lattice constants of the constituent binaries creating the alloy. For the system AlAs-GaAs, where both components have nearly the same interatomic spacing, the solid solution is very close to an ideal solution. For the system InP-InSb, however, where the two components are very different in size, the system exhibits a large deviation from ideality which gives rise to a miscibility gap, i.e., the two compounds do not form solid solutions over a certain composition range. One can easily realize that in order to gain definite information on the composition of the grown solid alloy when the composition of the gas phase (being a mixture of the reactants and the carrier gas) is known, thermodynamical estimations should be carried out.

The physical situation that occurs in the growth of an GaAs epilayer by MO CVD is illustrated in Fig. 2.12. The gaseous flow dynamics in an atmospheric or near atmospheric pressure reactor results in the formation of a boundary layer at the stationary substrate surface. The thickness of the boundary layer increases with lateral distance in the direction of flow. The decreased velocity of gases in this region results in a quasi-stagnant layer of gases that controls the transport of reactants to the substrate surface. The reactants are consumed as they reach the surface establishing at any point along the surface a steady state gradient that drives the diffusion of the reactants. The boundary layer is also characterized by a strong temperature gradient through which reactants diffuse and in which they may decompose to release the atomic species necessary for layer growth. If the rate of diffu-

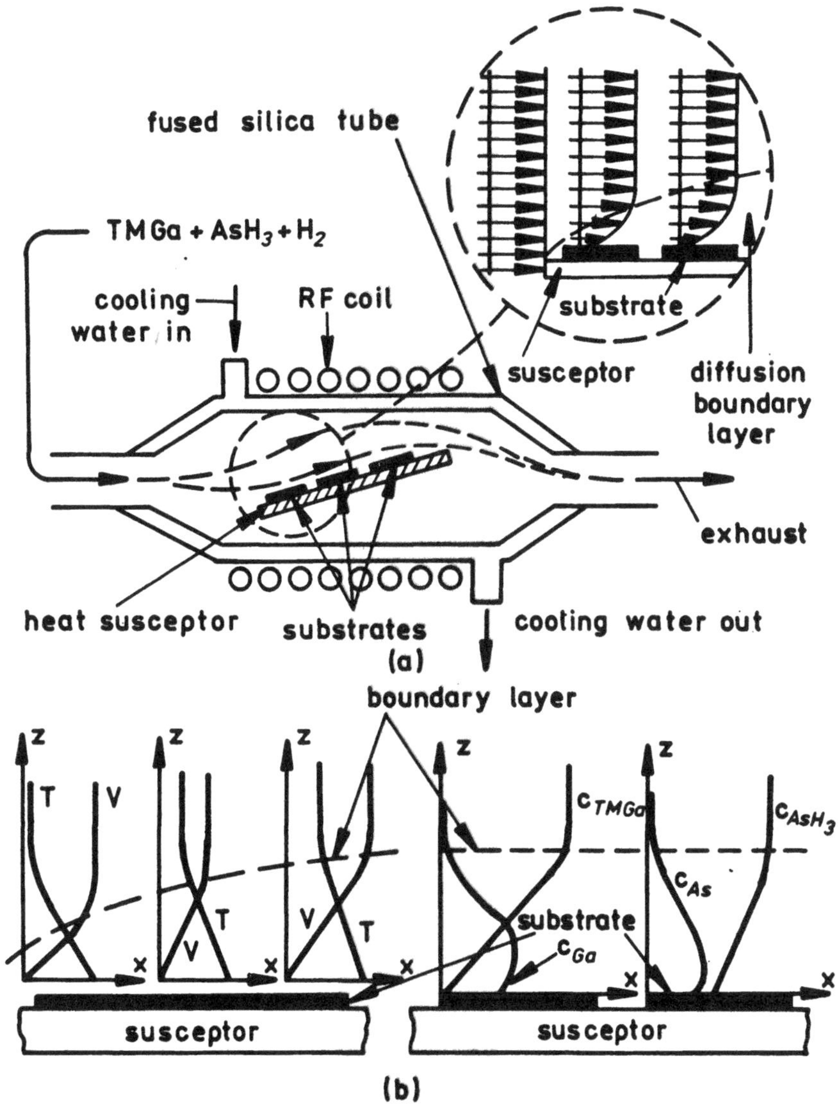

Fig. 2.12. a) Schematic drawing of the fused silica tube horizontal crystallization reactor for MO CVD cold wall technique, and b) the changes of temperature, flow velocity and reactants concentrations in the diffusion boundary layer.

sion through the boundary layer exceeds the rate of decomposition of the reactants then surface catalytic effects may become important in the film growth process. This seems likely in the case of GaAs growth using AsH_3 for the As source |18|.

The desired reaction which leads to crystal growth of a III-V compound is simply the decomposition of the adduct or individual group III and group V source molecules. This decomposition takes place either in the heated boundary layer through which the molecules must pass to reach the substrate, or on the substrate surface itself. For the growth of III-V compounds the input gas is highly supersaturated but equilibrium is established at the solid-vapour interface. This leads to a total depletion at the interface of one from the two components of the crystallized compound |74|.

Since the ratio of group V to group III elements partial pressures is large, the Ga partial pressure will be nearly zero at the substrate surface (see, Fig. 2.12b). The same situation will prevail for all group III elements at the interface during the growth of $A_xB_{1-x}C$ alloys with mixing on the group III sublattice. Thus, the group III distribution coefficient will be controlled by diffusion and since the group III diffusion coefficient D_{III} depends little upon the species, will be nearly unity for all of these systems. Moreover, since the ratio of group V to group III partial pressures in the input gas stream is high, the flux of the group III element through the boundary layer may limit the growth rate. This rate limiting flux may be written as |74|

$$I_{III} = D_{III}(p^o_{III} - p_{III})/(RT \cdot d_b) \quad , \qquad (2.2.7)$$

where d_b is the boundary layer thickness and p^o_{III} and p_{III} are the partial pressures of the group III element in the input gas stream and at the growing interface, respectively.

As it was discussed above, thermodynamic forces drive the epitaxial process. However, the real growth rate is often less than the thermodynamically allowed maximum growth rate. It means that the process is limited by physical kinetic processes occurring at the substrate surface. It is expected that the MO CVD kinetics is similar to those occurring in MBE growth. The kinetic processes in MO CVD are adsorption at the surface, surface reactions, attachment at an atomic step on the surface, etc. (see, Fig. 2.6). In principle, any of these steps may dominate the growth rate. The MO CVD process is limited by surface kinetics only at very low temperatures, hence, very little is known about the details of the surface processes in MO CVD. Usually, the slowest process in the series of steps involved in MO CVD is the diffusion of the reactants through the boundary layer, and hence this step is most often rate limiting in this growth process.

Doping during the growth of an epilayer is accomplished in the MO CVD process by introducing an appropriate gaseous reactant into the gas flow. It is expected that this reactant diffuses through the boundary layer in the same manner as the growth reactants do. The incorporation of the dopants into the crystal lattice may be controlled by either surface reactions (adsorption versus desorption or surface catalysis) or the termochemistry of the reactants. Presently, for doping of III-V compounds, organometallic reactants are typically used as p-type dopants (DEZn, Cp_2Mg) while hydrides (H_2Se, SiH_4) are used as n-type dopants |18|. In the case of organometallic dopant sources, surface kinetics controls the incorporation since the compounds decompose readily at the growth temperature. However, the doping efficiency of certain n-type dopants (particularly SiH_4) is controlled by the termochemistry of the hydride sources.

Concluding the considerations on the growth problems in MO CVD it is worth to summarize the effects of several growth variables on the growth rate of the epilayer |74|. Three cases of possible growth mechanisms in MO CVD should be considered here, in order to be general enough. These are: the growth limited by thermodynamics, the growth limited by diffusion of reactants in the boundary layer, and the growth limited by surface kinetics.

The gas flow velocity in the reactor, when keeping the partial pressures of the reactants fixed, affects the thermodynamically limited growth rate by passing more or less nutrients through the reactor. It also influences the boundary layer thickness and so affects also the growth rate in the diffusion limited case. An increase in flow velocity increases the growth rate of the epilayer in both cases.
Increasing the partial pressure of the reactants in the flowing gas stream increases the growth rate in all cases.

Increasing the substrate temperature increases the growth rate when it is limited by surface kinetic processes. However, for exothermic processes such as occurring in MO CVD, increasing the growth temperature decreases the extent of the reaction at equilibrium and so slows down the growth rate when limited by thermodynamics. Temperature has only a slight effect on the growth rate when the growth process is limited by diffusion. The effect of substrate orientations is, of course, only important when the growth is limited by surface kinetic processes.

2.2.3. Growth of superlattice structures

The growth of multilayer structures is accomplished in MO CVD by changing the gas composition in the reactor. The rate at which such a change takes place depends on total flow and the geometry of the reactor. At high flow rates, this exchange can be accomplished fast enough so that near-abrupt heterojunctions can be formed |18|. The time constant for the exchange is controlled by the total flow rate. Thus, the interface abruptness is also affected by the growth rate. This means that, e.g. in the case of III-V compounds, the interface abruptness is dependent on the concentration of group III reactants in the gas phase flowing through the reactor.

Experimental data concerning GaAs-$Al_xGa_{1-x}As$ superlattices |15| have provided evidence that MO CVD, especially when automated growth systems are used, may lead to superlattices of high structural quality with individual layers as thin as 1.5 nm. The thicknesses of the superlattice layers and of the transition widths at the heterointerfaces in these structures are usually observed by means of transmission electron microscopy. Such observations have confirmed |77| that it is possible to prepare with MO CVD atomically abrupt GaAs-AlAs heterojunction structures. Compositional changes from GaAs to AlAs realized over approximately one atomic layer have been ascertained. The surface roughness was also in this scale. Moreover, in GaAs-$Al_{0.55}Ga_{0.45}As$ heterostructures grown by MO CVD the GaAs layers were only 0.7 nm thick |77|.

These experimental results lead to a conclusion, that the MO CVD growth technique is competitive with MBE for the growth of thin epitaxial multilayer structures, at least when the GaAs-$Al_xGa_{1-x}As$

material system is considered.

As it has already been mentioned, in order to achieve minimal transition widths in the epilayer structures grown by MO CVD, the changes in gas composition presented to the substrate surface must have the steepest possible gradient. An initially perfect abrupt gas-composition change, brought about by opening or closing a valve at the inlet to the deposition zone of the reactor, can fade to a gradual change before arrival at the substrate surface. This produces a graded rather than an abrupt change of composition in the epitaxial layer structure |77|.

Let us discuss this problem qualitatively for the case in which the supply of a given component is switched on (switch-off effects can be treated along analogous lines). The following three factors reduce the intended infinite slope in the gas composition:

(i) After opening the valve which introduces the component, it takes time to build up the stationary concentration in the stagnant volume after the actual valve seal and connecting pipeline. In this part of the reactor system, relatively low flow rates are used. Consequently in the build-up phenomenon diffusion plays an important role leading to a gradual rather than a stepwise change in concentration.

(ii) The mentioned gradual front is transported by the carrier gas from a mixing chamber to the susceptor region. During this transport, diffusion processes will further reduce the slope of the concentration change.

(iii) When the front with its reduced slope has come close to the substrate, the boundary layer still has to be crossed for the components to reach the substrate surface itself. Transport through this barrier occurs by diffusion alone, which causes a further decrease of the slope.

The obvious conclusion which results from the above presented arguments is that the slope-decreasing factors have to be minimized in the optimized reactor design. A schematic drawing of such an optimized MO CVD reactor is shown in Fig. 2.13. Each component is switched in this reactor either to the mixing chamber or to a bypass line by means of a pair of air-operated bellows valves. The resistance for flow through both channels has been made equal. A single pressure line operates both inlet and bypass valves simultaneously, closing one valve while opening the other one.

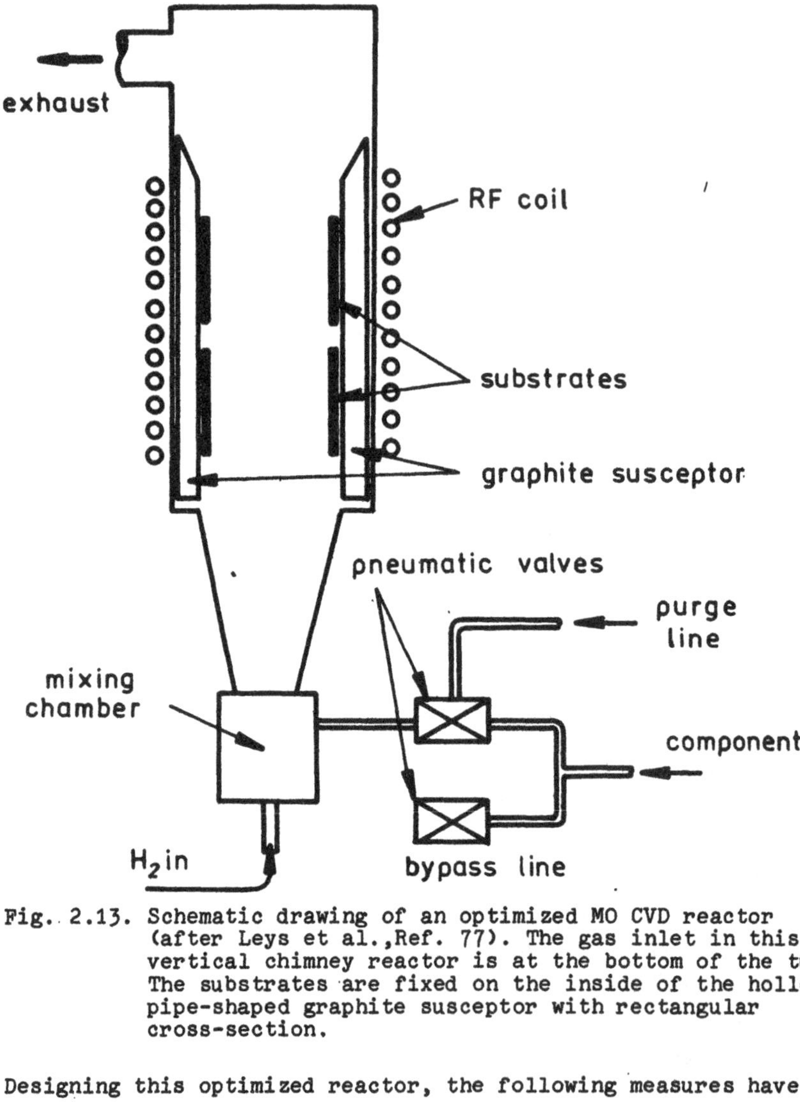

Fig. 2.13. Schematic drawing of an optimized MO CVD reactor (after Leys et al.,Ref. 77). The gas inlet in this vertical chimney reactor is at the bottom of the tube. The substrates are fixed on the inside of the hollow, pipe-shaped graphite susceptor with rectangular cross-section.

Designing this optimized reactor, the following measures have been taken to minimize the slope-decreasing factors |77|:

ad (i). The inlet valves were purged constantly by a hydrogen flow. This has largely eliminated the occurrence of stagnant volumes in the valves and connecting pipelines.

ad (ii). The transport time from the valve to the mixing chamber and from there to the substrate was kept minimal by minimizing the distances. The transport time includes extra time spent during passage through path-lengthening deviations from an ideal smooth streamline pattern. Such deviations, like gas recirculations in the reactor tube, have been minimized by two measures. Firstly, the reactor inlet was streamlined. Secondly, the disturbing effect of free convection has been avoided.

Free convection arises when the incoming cold gas approaches the hot susceptor region. In a horizontal reactor (see, Fig. 2.12) the free convection is perpendicular to the forced convection. This can produce serious disturbance of an otherwise smooth streamline pattern and favours gas recirculation. In the vertical chimney reactor shown in Fig. 2.13 the carrier gas with components is introduced at the bottom. The gas passes through a vertical hollow, pipe-shaped susceptor with rectangular cross section on whose inner walls the substrates are attached. With this geometry, the free convection (comparable to chimney draught) cooperates rather than interferes with the forced convection. Thus, a "natural" gas flow has been obtained.

ad (iii). The thickness of the boundary layer over the susceptor has been reduced by the cooperating free convection. Thus, the corresponding diffusion time to the substrate surface has been reduced as well.

The pipe-shaped susceptor of the reactor shown in Fig. 2.13 is made of graphite. It is heated by RF induction. The susceptor temperature, valve switching and mass flows are controlled by a microcomputer. Hydrogen, purified by means of a palladium diffusor is used as carrier gas at atmospheric pressure. Sources of TMGa, TMAl and AsH_3 are fitted. Growth of GaAs-AlAs and GaAs-$Al_xGa_{1-x}As$ super lattices was performed in this reactor with a susceptor temperature 700^oC. The total flow through the reactor tube was 7.5 l/min, the partial pressure of TMGa was $3\cdot10^{-5}$ atm, the arsine pressure was $7.5\cdot10^{-4}$ atm. Under these conditions the growth rate of GaAs was 0.35 nm/s. The partial pressure of TMAl could be varied from 0 to $1.2\cdot10^{-4}$ atm in order to grow $Al_xGa_{1-x}As$ layers with various compositions. Pure AlAs was grown with a TMAl partial pressure of $3.6\cdot10^{-5}$ atm |77|.

Despite the excellent results achieved in growing GaAs-$Al_xGa_{1-x}As$ superlattice structures by MO CVD when the described vertical chimney reactor was applied, some detailed discrepancies such as a nonlinear relationship between layer thickness and growth time, seen in very thin layers |76|, and the occurrence of non-planar interfaces during growth of pure AlAs layers |77|, raised doubts concerning the ideality of MO CVD grown materials in the direct vicinity to the heterojunction interfaces. Also investigations concerning laser emission and optical absorption in GaAs-$Al_{0.45}Ga_{0.55}As$ superlattices have evidenced material nonidealities existing at the heterointerfaces. The observed spectral shift of the laser emission line towards lower energies, which indicates that the superlattice quantum wells exhibit an effective size larger than the intended size should be according to the performed MO CVD growth process, could be understood only by assuming that both Al-rich and Al-depleted aggregates (clusters) exist at the heterointerfaces on the side of the barrier layers of these superlattices |14|. Interface anomalies have also been observed with ellipsometry |78|,|79| when $Al_xGa_{1-x}As$ was grown on GaAs. However, GaAs growth on $Al_xGa_{1-x}As$ yields very sharp heterojunctions.

These listed sxperimental facts do not mean, in any way, that the MO CVD technique exhibits inherent shortcomings which cannot be eliminated, when greater effort will be exerted on investigations concerning the growth peculiarities characteristic for this crystallization technique.

2.3. Growth peculiarities related to different types of superlattices

Based on the discussion of fundamental growth processes and results obtained to date, the conclusion is evident that both techniques, MBE and MO CVD, are well suited for growing superlattices of the GaAs-$Al_xGa_{1-x}As$ material system. What is less clear is the specific role each of these techniques will play in the future.

MBE seems certain to dominate the fundamental superlattice structures research |80-86| until more understanding of MO CVD growth processes is achieved. Similarly, MO CVD will probably dominate the device development research and production. The degree to which MO CVD is able to penetrate the field of electronic devices with superlattice structures (see, Chapter 5) will depend upon the solution of certain problems that currently limit the performance of MO CVD grown superlattice structures. Devices with superlattices based on InP-InGaAs and InP-InGaAsP material systems will most likely be fabricated by MO CVD, owing to the difficulties of growing phosphorus containing compounds by MBE. However, progress recently made in this field with MBE |8|, may change this opinion soon. It is evident that the development of both, the MBE and the MO CVD techniques continues unabated, fueled by the ingenuity of materials researchers and the competitive presence of the other technique.

So far, the technological problems related to the preparation of superlattice structures have been exemplified by the growth peculiarities of III-V compounds. Among these compounds the GaAs-Al_x $Ga_{1-x}As$ material system has been analyzed most frequently. It is, however, obvious that for other materials used to obtain superlattices the growth problems may be quite different. Therefore, to complete this technological chapter, a short survey of these problems related to both, compositional and doping superlattices, seems to be indispensable.

2.3.1. Growth of II-VI and IV-VI compound compositional superlattices

Mercury cadmium telluride is the most important material for infrared detectors and imaging arrays |87|. The energy gap of this compound is a function of the mole fraction x of CdTe and the ambient temperature. It ranges from -0.300 eV for semimetallic HgTe to 1.648 eV for semiconductive CdTe at 4.2 K temperature. The variable band gap aspect and high electron mobilities make this material nearly ideal for device applications. The need for large--area $Hg_{1-x}Cd_xTe$ material of good crystalline quality for detectors and advanced focal plane arrays has given rise to growing interest toward heterostructures and superlattices made of this material system.
The HgTe-CdTe superlattice is a limiting case of a multiheterostructure of $Hg_{1-x}Cd_xTe$-CdTe with x=0. The most suitable growth technique for these superlattices is the MBE technique. The best HgTe-CdTe superlattices are grown by MBE at temperatures as low as 473 K. This fact allows to avoid any serious interdiffusion during the growth and ensures the required abruptness of the super-

attice heterointerfaces. From Auger electron spectroscopy profiling, the HgTe-CdTe interface was found to be less than 4 nm. From infrared magneto-optics experiments the interdiffusion width in a superlattice grown at 473 K appeared to be less than 2.5 nm. Preliminary results from ellipsometry measurements suggest that this interdiffusion width could be in the range of 1 nm |33|. Such results can only be obtained with the MBE technique, which will remain the most suitable technique for crystallization of HgTe-CdTe superlattices until MO CVD low temperature technology will be developed.
The MBE grown HgTe and CdTe layers of the superlattice possess sufficiently high structural perfection, which may still be improved in the near future |33|,|88|. RHEED investigations have shown that the growth temperature plays an important role in MBE of this superlattice. Like for single epitaxial layers, as the substrate temperature is increased, e.g. from 390 to 473 K, a crystal perfection improvement occurs for both CdTe and HgTe layers. However, below 473 K the epitaxy occurs with more difficulty for CdTe than for HgTe, and the HgTe crystal quality seems better than the CdTe.

So far, HgTe-CdTe superlattices have been grown on CdTe (111) substrates by using three different effusion cells |87|. Since the sticking coefficient of Hg decreases sharply as temperature is increased, a metallic Hg vapour source rather than a HgTe compound source is preferred to yield a sufficiently high flux of Hg. Thus, elemental Hg and Te are usually used as sources to grow HgTe layers and a single CdTe source to grow CdTe. The Hg cell applied is most frequently transferable and is introduced into the growth chamber only during the actual deposition in order to reduce the pressure in the chamber. With such a cell the Hg flux of 10^{18} atoms $cm^{-2}s^{-1}$ can be maintained for several hours at the optimal growth temperature. The deposition rates for both CdTe and HgTe layers are held constant at about 2 monolayers/s during the superlattice growth process.
Under these conditions a 150-period superlattice consisting alternately of HgTe, 4 nm thick, and CdTe, 6 nm thick, layers could be repetitively grown |33|.

Another interesting CdTe-based superlattice material system uses $Cd_{1-x}Mn_xTe$, a semimagnetic semiconductor compound. The band gap of

this compound increases with x, i.e., with increasing Mn content. As a consequence $Cd_{1-x}Mn_xTe$ layers are appropriate for use as barrier layers between CdTe quantum wells. It is well known |89| that the presence of Mn ions in the crystal lattice of the bulk semimagnetic semiconductor leads to a spin-spin exchange interaction between the localized magnetic moments associated with the Mn ions and the conduction electrons. This interaction gives rise to an extremely large Faraday rotation, giant negative magnetoresistance, and large positive electronic g factors. The $CdTe-Cd_{1-x}Mn_xTe$ superlattice structures offer an unique possibility to study these fundamental magnetic interactions or magneto-optic phenomena in two-dimensional structures of the semimagnetic semiconductor. These superlattices may be also exploited for increasing the applications potential of CdTe, which creates there near ideal square-shaped quantum wells. The heterointerfaces are extremely sharp, because no interdiffusion of Mn or other grading effects occur in the interface region |90|.
Superlattice structures of $Cd_{1-x}Mn_xTe$ ($x = 0.06 - 0.50$) alternating with CdTe have been successfully grown with the MBE technique |91-94| on several substrates ($GaAs, Al_2O_3, CdTe$). However, when the substrate used is different than CdTe, first before the superlattice starts to be grown, a buffer layer of CdTe (2-5 μm thick) is grown. These buffer layers exhibit usually the (111) crystallographic orientation even then, when (100) oriented GaAs |95| or (0001) oriented Al_2O_3 |93| substrates are used. Therefore, the superlattice layers are also (111) oriented.

The MBE growth is usually carried out with substrate temperature equal to about 580 K at the rates of 0.1 to 0.2 nm/s. The molecular beams are generated in the simplest case by two effusion cells, one containing vacuum-distilled Mn (heated to about 1120 K) and the other one containing CdTe of 6N purity (heated to about 600K). The growth conditions specified above allowed to grow high quality 240-period $CdTe-Cd_{1-x}Mn_xTe$ superlattices consisting of CdTe quantum well layers only 1.6 nm thick sandwiched between $Cd_{0.75}Mn_{0.25}Te$ barrier layers 2.1 nm thick |93|. The high quality of these superlattices has been evidenced not only by x-ray diffraction measurements, but also by an extremely bright photoluminescence spectrum. The luminescence radiation emitted by the superlattice was two orders of magnitude brighter than usually observed in bulk hydroplane polished CdTe or CdTe on sapphire epitaxial films. This enhancement in the photoluminescence efficiency which results from the localization of excitons at the heterointerfaces |96|, may have important implications concerning the use of this superlattices in future optoelectronic and integrated optics applications.

The last material system belonging to the II-VI group, which has been applied for growing superlattice structures, is the ZnS-ZnSe system |28|. Superlattices made of these materials have been grown on (100) oriented GaAs substrates by hot wall epitaxy in a system providing two independent evaporation crucibles, one for ZnS and the other one for ZnSe (see, Fig. 2.14). These superlattices have consisted of more than 150 strained layers of thicknesses of 1 nm for ZnSe and 4.9 nm for ZnS, in the best structures grown. Until now, however, little is known about technological peculiarities related to this superlattice material system.

Superlattices made of IV-VI compound semiconductors such as PbTe, SnTe, $Pb_{1-x}Sn_xTe$ etc. represent quite special systems. This results from such physical properties of the constituent materials like narrow energy gaps, high carrier mobilities and large static dielectric constants associated with a ferroelectric lattice instability. The most intensively studied superlattice representing this material group is the PbTe-$Pb_{1-x}Sn_xTe$ superlattice. This superlattice is usually grown by a modified hot wall technique |27|, |97| with layer thicknesses ranging from 10 to about 50 nm, on cleaved (111) oriented BaF_2 substrates.

The growth conditions are defined by the temperatures of the two evaporating sources, the hot wall tube and the substrate (see, Fig. 2.14). At typical temperatures equal to: 800 K for the sources, 770-820 K for the substrate, and 820 K for the wall, the growth process performed in a vacuum environment of about 10^{-6} torr leads to PbTe layers which exhibit n-type conductivity, and to $Pb_{1-x}Sn_xTe$ layers exhibiting p-type conductivity. The typical concentrations of free charge carriers are equal to: n = (0.5-3) 10^{17} cm^{-3} in PbTe and p = (2-10)10^{17} cm^{-3} in $Pb_{1-x}Sn_xTe$. Before the superlattice structure started to be grown, a PbTe buffer layer about 500 nm thick is usually grown on the BaF_2 substrate. Layer thicknesses are selected by two limitations |98|: (i) the lower boundary is limited by the composition gradients due to in-

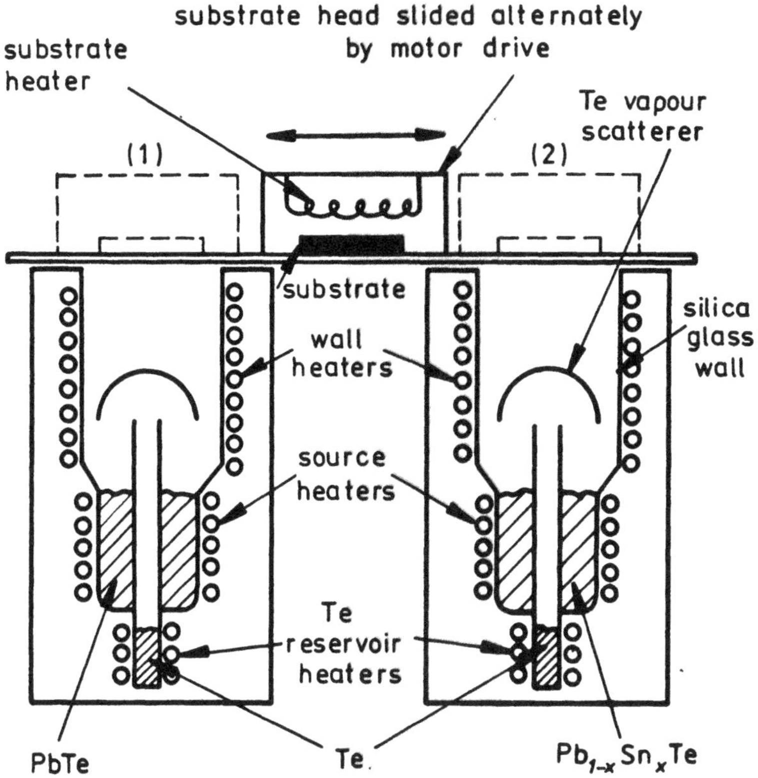

Fig. 2.14. Schematic drawing of the hot wall epitaxy system used to prepare PbTe-PbSnTe superlattices (after Kinoshita et al., Ref. 97).
The whole system is placed in a vacuum of about 10^{-6} torr. Te charge in the reservoirs is utilized when growth of p-type PbTe and p-type PbSnTe is required. Excess amount of Te in the epilayer induces deviations from stoichiometry. When growing ZnS-ZnSe superlattices, the source crucibles are filled with Zn in both furnaces, and the reservoirs are filled with Se in the first furnace and with S in the second.

terdiffusion of the constituents, (ii) the upper is determined by the mean free path of the free carriers to achieve a transition from superlattice to bulk behaviour.

Recently $PbTe-Pb_{1-x}Sn_xTe$ superlattices have been grown on (100)-oriented KCl substrates. The reason for selecting this substrate instead of BaF_2(111) is related to the fact that in (111) oriented PbSnTe layers two kinds of energetic subbands with light and heavy effective masses occur. This complicates the features of the electronic states in the superlattices. Preparation of a PbTe-PbSnTe superlattice on a (100)-oriented KCl substrate is believed to reduce this complication and to enable the application of such superlattices to PbSnTe quantum well lasers |29|.

2.3.2. Growth of doping III-V and IV-VI compound superlattices

The most useful growth technique for III-V compound doping superlattices is MBE |99|, while for growing of doping superlattices made of PbTe and related compounds hot wall epitaxy (HWE) is most frequently used |100|. The incorporation of dopants in controlled amounts into the MBE grown superlattice structures is achieved by using additional effusion cells which contain the appropriate doping elements. Doping during the HWE growth process can be achieved by creating in this process intentional non-stoichiometry of the grown epilayers. In the case of PbTe-based superlattices concentrations of n-and p-type carriers up to 10^{18} cm^{-3} in the layers of the host crystal may be easily reached. Each Pb vacancy forms a double negatively charged deffect whose charge is neutralized by two holes in the valence band. The Te vacancies behave correspondingly, generating a double positively charged defect and two free electrons in the conduction band. Thus, PbTe grown with Te in excess will exhibit p-type conductivity, while PbTe grown with Pb in excess will be of the n-type.
In order to be more precise, we will restrict here our considerations only to GaAs doping superlattices, as a representative for the III-V compounds, and to PbTe doping superlattices, as a representative for the IV-VI compounds.
In the case of GaAs, only the application of Si as donor and Be as acceptor impurities yields a uniform doping within each individual layer and a controlled thickness with constant periodicity also for a large number of layers in sequence |99|. Only in this case the transition regions, i.e. the "interfaces" between the individual GaAs layers remain extremely smooth down to atomic

steps also over large substrate areas |101|.

Typically, GaAs doping superlattices are grown on semi-insulating (100)-oriented GaAs substrates at 823 K. The n- and p-doped layers of these superlattices are of equal thicknesses $d_n = d_p = 25$ nm, and of equal doping level, with donor and acceptor concentrations $N_D = N_A = 1\cdot 10^{18}\,cm^{-3}$. For such design parameters, the amplitude V_{Do} of the periodic space charge potential in the non-excited state is equal to 0.113 ev |102|. However, the band structure of GaAs doping superlattices in the ground state may be tailored by the appropriate choice of the design parameters. According to the Eq. (1.3.1) the periodic space-charge potential is determined by the doping concentration and by the thickness of the constituent layers. Therefore, different types of superlattices may be produced intentionally by preselecting their design parameters.

Let us consider, as an example, the following three types of GaAs doping superlattices |99|, the real-space energy band diagrams of which are shown in Fig. 2.15

(i) In the case of low doping concentrations and a small superlattice period d, twice the space-charge potential amplitude $2V_{Do}$ is smaller than the band gap of the homogeneous bulk material. With equal doping densities in the constituent n and p layers, i.e. $N_D d_n = N_A d_p$, all excess electrons of the donors recombine with the holes of the acceptors. This structure represents a compensated intrinsic superlattice having no free carriers and with the Fermi level E_F in the middle of the effective energy gap (see, Fig. 2.15a).

(ii) When keeping the value of $N_A d_p$ as before, and increasing simultaneously the value of $N_D d_n$, one obtains the n-type superlattice, having the Fermi level in the conduction band (see, Fig.2. 15b). Now the n-layers contain excess electrons of density $n = N_D d_n - N_A d_p$, whereas the p-layers are totally depleted. It should be noticed that the superlattice will exhibit the p-type conductivity as a whole, if $N_A d_p > N_D d_p$.

(iii) If the values of $N_D d_n$ and $N_A d_p$ are large, the amplitude $2V_{Do}$ reaches its maximum value which is slightly more than the band gap of the homogeneous bulk (see, Fig. 2.15c). In the ground state there are now free electrons in the n layers and free holes in the p layers. The superlattice resembles a semimetal, which may be, however, n-type or p-type, depending on the choice of the doping density in the respective layer type.

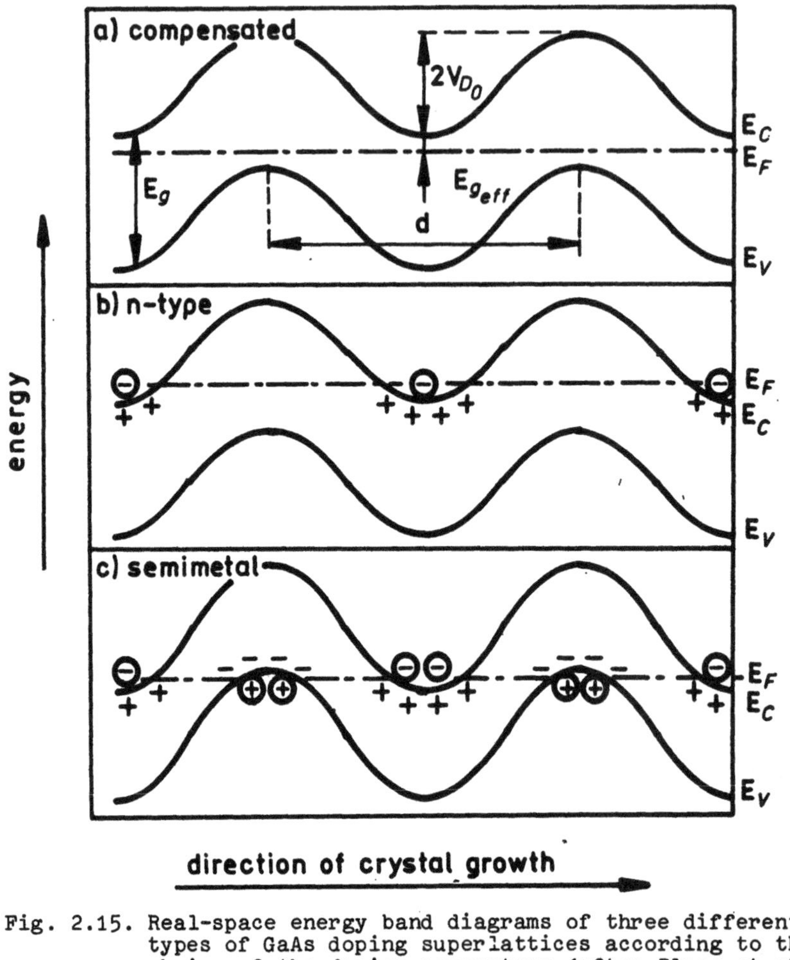

Fig. 2.15. Real-space energy band diagrams of three different types of GaAs doping superlattices according to the choice of the design parameters (after Ploog et al., Ref.99).
a) compensated intrinsic type ($2V_{Do} < E_g, N_D d_n = N_A d_p$)
b) n-type ($2V_{Do} < E_g$, $N_D d_n > N_A d_p$)
c) semimetal n-type ($2V_{Do} > E_g$)

A PbTe doping superlattice can be grown by periodic n- and p-doping just as in the case of GaAs. The electronic structure, however, becomes now quite different due to the different material parameters of the host crystal. From Eq. (1.3.1) it follows that the superlattice period and the doping concentration in PbTe doping superlattices have to be considerably larger than in their III-V counterparts in order to obtain values of V_{Do} which are comparable with the band gap E_g of the host material. The reason for that is the static dielectric constant $\varepsilon \approx 1500$ (at T = 0 K) of PbTe, which is by more than two orders of magnitude larger than that of GaAs. Therefore, the situation similar to that which is shown in Fig. 2.15a, but with a modulation of the band-edges such that the effective band gap in the ground state of the superlattice nearly vanishes, can be realized only with individual layers as thick as $d_n = d_p = 200$ nm if $2N_D = 2N_A = 10^{18}$ cm^{-3} |100|.

The first PbTe doping superlattices have been grown in 1981 |103| however, little detailed information is available at present on the growth peculiarities of these structures.

In conclusion of these considerations related to doping superlattices, it should be mentioned that a severe diffusion erosion may occur in these superlattices, when exposing them to elevated temperatures |104|. Therefore, care must be taken during the growth or during the operation in a device structure to avoid the disturbing diffusion effects.

2.3.3. Growth of silicon-based superlattices

Three types of superlattices can be prepared with silicon or silicon-based materials. The first is the strained-layer type-I compositional superlattice of Ge_xSi_{1-x}/Si system. The second is the Si-doping superlattice , and the third is the amorphous superlattice made of one of the following material systems: a-Si:H/a-Ge:H, a-Si:H/a-SiN_x:H or a-Si:H/$Si_{1-x}C_c$:H. Although representative superlattice structures have already been grown for almost all of the mentioned material systems, the investigations related to the growth processes are at present in their infancy. Therefore, only some preliminary data can be given here in order to exemplify the growth peculiarities of these interesting superlattice structures.

The Ge_xSi_{1-x}/Si superlattices are usually grown by MBE on (100)-oriented Si substrates at 823 K |105|. This growth temperature is sufficiently low for an effective inhibition of the dislocation formation in the lattice mismatched epilayers, even at layer thicknesses up to 250 nm, which is larger than the critical thickness for the formation of dislocations. It has been shown |106| that the Ge_xSi_{1-x} alloys deposited By MBE on silicon can be prepared as metastable coherently strained layers. These strained layers can, however, thermally relax by generating a set of dislocations near the interface, when annealed at about 1020 K for 30 min |107|. Initial relaxation of films 50 nm or thicker proceeds by logarithmic time-dependent growth of dislocations near the interface, a creep process which partially relieves local strain and has comparatively less influence on the strain at the surface. Relaxation of macroscopic strain is associated with a subsequent stage of Ge diffusion, particularly in films less than 10 nm thick. It has also been shown |108| that the relaxation of the Ge_xSi_{1-x}/Si superlattice can significantly affect lattice parameter measurements in this strained layer material system.

The Si doping superlattices described theoretically in detail |109| will be grown probably with MBE |4|. The most important growth problem will be the precise control of the alternating doping during the MBE process. There are three ways to dope the growing Si layer in the MBE process: (i) by co-evaporation of doping material, (ii) by dopant transfer from a doped silicon source, and (iii) by simultaneous low energy ion implantation during the growth |4|.

Co-evaporation of doping material during MBE growth is commonly used to dope III-V compound films, and this technique seems to be the most probable for being applied to the preparation of Si doping superlattices. For n-type doping of Si grown by MBE antimony has been used because it has a lower vapour pressure and chemical reactivity than arsenic or phosphorus, and as a consequence is easier to handle. For p-type doping gallium, aluminium and indium have been used, however, only doping with Ga leads to impurity profiles in the Si film, which can be controlled accurately for substrate temperatures in the range from 873 K to 1073 K, and for carrier densities in the range from 10^{14} to $5 \cdot 10^{17}$ cm^{-3}, by changing the temperature of the effusion cell containing the dopant species |4|.

Amorphous silicon-based superlattices have been grown by a plasma-assisted CVD process in which the composition of the reactive gases is changed periodically in the reaction chamber |110|. In the a-Si:H/a-SiN_x:H superlattice the silicon layers have been made from pure SiH_4, while the silicon nitride layers have been made from 20 vol% SiH_4 and 80 vol% NH_3. The residence time of the gas in the reactor ($\approx$1 s) was short compared to the time required to grow a monolayer ($\approx$3 s). Thus, the gas could be exchanged rapidly enough to achieve sharp interfaces between layers. The films were deposited on quartz substrates held at 493 K on a 6-in.-diam. anode in an RF (13.56 MHz,5W) capacitive-geometry reactor. The plasma was maintained continuously while the gases were changed. Under the described conditions well-defined layers for a 6.8 nm period a-Si:H/a-SiN_x:H superlattice have been grown |110|.

The fact that extremely smooth (RMS roughness $\approx$ 0.5 nm), parallel and uniform layers with atomically sharp interfaces are achievable also for the other silicon-based amorphous superlattices, has been demonstrated by transmission electron microscopy and X-ray diffraction measurements |111|.

2.4. Literature to chapter 2

|1| Madhukar,A.: Modulated semiconductor structures-an overview of some basic considerations for growth and desired electronic structure, J.Vac.Sci.Technol. 20 (1982)2, 149-161

|2| Esaki,L., and L.L.Chang: Semiconductor superfine structures by computer-controlled molecular beam epitaxy, Thin Solid Films 36 (1976) 1, 285-298

|3| Cho,A.Y.: Growth of III-V semiconductors by molecular beam epitaxy and their properties, Thin Solid Films 100 (1983) 1, 291-317

|4| Ota,Y: Silicon molecular beam epitaxy, Thin Solid Films 106 (1983) 1/2, 1-136

|5| Thomas,G.E., L.J.Beckers, J.J.Vrakking, and B.R. de Koning: Ion beam epiplanation, J.Crystal Growth 56 (1982) 2, 557-575

|6| Kawabe,M., M.Kondo, N.Matsuura, and K.Yamamoto: Photoluminescence of $Al_xGa_{1-x}As/Al_yGa_{1-y}As$ multiquantum wells grown by pulsed molecular beam epitaxy, Jap.J.Appl.Phys. 22 (1983) 2, L64-L66

|7| Herman,M.A., O.Jylhä, and M.Pessa: Atomic layer epitaxy of $Cd_{1-x}Mn_xTe$ grown on CdTe (111) B substrates, J.Crystal Growth 66 (1984) 2, 480-483

|8| Panish,M.B., H.Temkin, and S.Sumski: Gas source MBE of InP and $Ga_xIn_{1-x}P_yAs_{1-y}$ - materials properties and heterostructure lasers, J.Vac.Sci.Technol. B3 (1985) 2, 657-665

|9| Sakamoto,T., H.Funabashi, K.Ohta, T.Nakagawa, N.J.Kawai, T. Kojima, and Y.Bando: Well defined superlattice structures made by phase-locked epitaxy using RHEED intensity oscillations, Superlattices and Microstructures 1 (1985) 4, 347-352

|10| Heckingbottom,R., G.J.Davies, and K.A.Prior: Growth and doping of GaAs using MBE-thermodynamic and kinetic aspects, Surface Sci. 132 (1983) 1, 375-389

|11| Madhukar,A.: Far from equilibrium vapour phase growth of lattice matched III-V compound semiconductor interfaces-some basic concepts and Monte-Carlo computer simulations, Surface Sci. 132 (1983) 1, 344-374

|12| Dupuis,R.D.:Metalorganic chemical vapour deposition of III -V semiconductors, Science 226 (1984) 4675, 623-629

|13| Dupuis,R.D.: AlGaAs-GaAs layers grown by metalorganic chemical vapour deposition-a review, J.Crystal Growth 55 (1981) 1, 213-222

|14| Coleman,J.J., P.D.Dapkus, M.D.Camras, N.Holonyak, Jr., W.D. Laidig, T.S.Low, M.S.Burroughs, and K.Hess: Absorption, simulated emission, and clustering in AlAs-$Al_xGa_{1-x}As$-GaAs superlattices, J.Appl.Phys. 52 (1981) 12, 7291-72-95

|15| Griffiths,R.J.M., N.G.Chew, A.G.Cullis, and G.C.Joyce: Structure of GaAs-$Al_xGa_{1-x}As$ superlattices grown by MO CVD, Electronics Lett. 19 (1983) 23, 988-989

|16| Soga,T., S.Hattori, S.Sakai, M.Takeyasu, and M.Umeno: Characterisation of epitaxially grown GaAs on Si substrates with III-V compounds intermediate layers by MO CVD, J.Appl.Phys. 57 (1985) 10, 4578-4582

|17| Balk,P., and E.Veuhoff: Deposition of III-V compounds by MO CVD and in halogen transport systems-a critical comparison, J.Crystal Growth 55 (1981) 1, 35-41

|18| Dapkus,P.D.: A critical comparison of MO CVD and MBE for heterojunction devices, J.Crystal Growth 68 (1984) 1,345-355

|19| Manasevit,H.M.: Recollections and reflections of MO CVD, J.Crystal Growth 55 (1981) 1, 1-9

|20| Moss,R.H.: Adducts in the growth of III-V compounds, J.Crystal Growth 68 (1984) 1, 78-87

|21| Elwell,D., and H.J.Scheel: Crystal growth from high-temperature solutions, chap.8, Academic Press, London 1975.

|22| Ufimcev,V.B., and R.C.Akchurin: Fiziko-khimicheskie osnovy zhidkofaznoi epitaksji, Izd.Metalurgija, Moscow 1983.

|23| Woodall,J.M.: Solution grown $Ga_{1-x}Al_xAs$ superlattice structures, J.Crystal Growth 12 (1972) 1, 32-38

|24| Moon,R.L.: Liquid phase epitaxy, in: Crystal growth, 2nd ed., pp.421-461, B.R.Pamplin (ed.), Pergamon Press, Oxford 1980.

|25| Benz,K.W., and E.Bauser: Growth of binary III-V semiconductors from metallic solutions, in: Crystals-growth, properties, and applications v.3,1-48, H.C.Freyhardt (ed.), Springer-Verlag, Heidelberg 1980.

|26| Lopez-Otero,A.: Hot wall epitaxy, Thin Solid Films 49 (1978) 1, 3-57

|27| Clemens,H., E.J.Fantner, and G.Bauer: Hot wall epitaxy system for the growth of multilayer IV-VI compound heterostructures, Rev.Sci.Instrum. 54 (1983) 6, 685-689

|28| Fujiyasu,H., H.Takahashi, H.Shimizu, A.Sasaki, and H.Kuwabara: Optical properties of ZnS-ZnSe superlattices prepared by a HWE, Proceedings of the 17th.Intern.Conf.on Physics of Semiconductors, San Francisco, 1984

|29| Fujiyasu,H., A.Ishida, H.Kuwabara, S.Shimomura, S.Takaoka, and K.Murase: Optical and electrical properties of $PbTe-Pb_{1-x}Sn_xTe$ superlattices prepared on KCl by a HWE, Surface Sci. 142 (1984) 1/2, 579-585

|30| Norris,P., J.Black, S.Zemon, and G.Lambert: Reduced pressure MO VPE growth and characterization of GaAs/AlGaAs heterostructures using triethylgallium source, J.Crystal Growth 68 (1984) 1, 437-444

|31| Ploog,K.: MBE of III-V compounds, in: Crystals-growth, properties, and applications v.3, 73-162, H.C. Freyhadt (ed.) Springer-Verlag, Heidelberg 1980.

|32| Foxon,C.T., and B.A.Joyce: Fundamental aspects of MBE, in: Current topics in material science v.7, 1-68, E. Kaldis (ed.), North-Holland P.C. 1981.

|33| Faurie,J.P., M.Boukerche, J.Reno, S.Sivananthan, and C.Hsu: MBE of alloys and superlattices involving mercury, J.Vac.Sci Technol. A3 (1985) 1, 55-59

|34| Herman,M.A.: Physical problems concerning effusion processes of semiconductors in MBE, Vacuum 32 (1982) 9, 555-565

|35| Svensson,S.P., and T.G.Andersson: Film thickness distribution at oblique evaporation, J.Vac.Sci.Technol. 20 (1982) 2, 245-247

|36| Myers,T.H., and J.F.Schetzina: Molecular beam source for high vapour pressure materials, J.Vac.Sci.Technol. 20 (1982) 2, 134-136

|37| Curless, J.A.: MBE beam flux modeling, J.Vac.Sci.Technol. B3 (1985) 2, 531-534

|38| Stanchak,C.M., H.Morkoc, L.C.Witkowski, and T.J.Drummond: Automatic shutter controller for MBE, Rev.Sci.Instrum. 52 (1981) 3, 438-442

|39| Fischer,A., K.Graf,M.Hafendörfer, H.Künzel,and K.Ploog: Rechnersteuerung für den Betrieb von MBE Anlagen, Technisches Messen 49 (1982) 11, 403-408, ibid 49 (1982) 12, 461-470

|40| Shchekochikhin,Yu,M.: Mathematical modelling of the multicomponent film forming processes from molecular beams, in: Proceedings of the Summer School on Physical and Technical Basis of MBE, Eberswalde GDR p. 86-99, 1979

|41| Svensson, S.P., and T.G.Andersson: Transformation between perpendicular and oblique incidence at evaporation, J.Phys.E: Sci.Instrum. 14 (1981)3, 1076-1077

|42| Collins,D.M.: On the use of "downward-looking" sources in MBE systems, J.Vac.Sci.Technol. 20 (1982) 2, 250-251

|43| Bosacchi,A., S.Franchi, P.Allegri, and V.Avanzini: MBE evaporation source fitted with shutter and water-cooled jacket, J.Vac.Sci.Technol. 21 (1982) 3, 897-898

|44| Kubiak,R.A., P.Driscoll, and E.H.C.Parker: A simple source cell design for MBE, J.Vac.Sci.Technol. 20 (1982) 2, 252-253

|45| Shen,L.Y.L.: Angular distribution of Molecular beams from modified Knudsen cells for MBE, J.Vac.Sci.Technol. 15 (1978) 1, 10-12

|46| Clausing,P.: Über die strahlformung bei der Molekularströmung, Zeitschrift für Physik, 66 (1930) 7/8, 471-476

|47| Clausing,P.: Über die Strömung sehr verdünnter Gase durch Röhren von beliebiger Länge, Annalen der Physik, 5 Folge, 12 (1932) 2, 961-989

|48| Lewis,B., and J.C.Anderson: Nucleation and growth of thin films, Academic Press, New York 1978.

|49| Venables,J.A., G.D.T.Spiller, M.Hanbücken: Nucleation and growth of thin films, Rep.Progr.Phys. 47 (1984) 2, 399-459

|50| Schmidt,L.D.: Condensation kinetics and mechanisms, in: The Physical basis for heterogeneous catalysis, E.Dräuglis, and R.I.Jaffee (eds.), Plenum Press, New York,1975, pp.451-465

|51| Ehrlich,G.: The mechanism of chemisorption on metals, J.Phys. ChemSolids 1 (1956) 1,3-13, ibid. 5 (1958) 1, 47-77

|52| Kisliuk,P.: The sticking probabilities of gases chemisorbed on the surfaces of solids, J.Phys.Chem.Solids 3 (1957) 1, 95-101, ibid. 5 (1958) 1,78-84

|53| Czanderna,A.W. (ed.): Methods of surface analysis, Elsevier Sc.P.C. Amsterdam 1975.

|54| Foxon,C.T., and B.A.Joyce: Interaction kinetics of As_2 and Ga on {100} GaAs surfaces, Surface Sci. 64 (1977) 1,293-304

|55| Foxon,C.T., M.R.Boudry, and B.A.Joyce: Evaluation of surface kinetic data by the transform analysis of modulated molecular beam measurements, Surface Sci. 44 (1974) 1, 69-92

|56| Foxon,C.T., and B.A.Joyce: Interaction kinetics of As_4 and Ga on {100} GaAs surfaces using a modulated molecular beam technique, Surface Sci. 50 (1975) 2, 434-450

|57| Kurz,E.A., and J.B.Hudson: Improvements to the molecular beam Fourier analysis technique, Appl.Surface Sci. 17 (1984) 2, 485-496

|58| Foxon,C.T.: Molecular beam epitaxy, Acta Electronica 21 (1978) 2, 139-150

|59| Foxon,C.T., and B.A.Joyce: Surface processes controlling the growth of $Ga_xIn_{1-x}As$ and $Ga_xIn_{1-x}P$ alloy films by MBE, J. Crystal Growth 44 (1978) 1, 75-83

|60| Singh,J., and K.K.Bajaj: Theoretical studies of the intrinsic quality of GaAs/AlGaAs interfaces grown by MBE-role of kinetic processes, J.Vac.Sci.Technol. B3 (1985) 2, 520-523

|61| Larsen,P.K., J.H.Neave, J.F. van der Veen, P.J.Dobson, and B.A.Joyce: GaAs(001) - c(4x4): A chemisorbed structure,Phys. Rev. B 27 (1983) 8, 4966-4977

|62| Tong,S.Y., W.N.Mei, and G.Xu: The geometric structure of the GaAs(111) and (110) surfaces, J.Vac.Sci.Technol. B2 (1984) 3, 393-398

|63| Veen van der, J.F., P.K.Larsen, J.H.Neave, and B.A.Joyce: The GaAs(001)-c(4x4) and (2x4) reconstructions: A comparative photoemission study, Solid State Commun. 49 (1984) 7,659-662

|64| Lievin,J.L., and F.Alexandre: Ultra-high doping levels of GaAs with beryllium by MBE, Electronics Lett. 21 (1985) 10, 413-414

|65| Harris,J.J., D.E.Ashenford, C.T.Foxon, P.J.Dobson, and B.A. Joyce: Kinetic limitations to surface segregation during MBE growth of III-V compounds: Sn in GaAs, Appl.Phys. A33 (1984) 1, 87-92

|66| Chai,Y.G., R.Chow, and C.E.C.Wood: The effect of growth conditions on Si incorporation in MBE GaAs, Appl.Phys.Lett. 39 (1981) 10, 800-803

|67| Salmon,L.G., and I.J.D'Haenens: The effect of aluminum composition on silicon donor behaviour in $Al_xGa_{1-x}As$, J.Vac.Sci Technol. B2 (1984) 2, 197-200

|68| Chand,N., R.Fischer, J.Klem, T.Henderson, P.Pearah, W.T. Masselink, Y.C.Chang, and H.Morkoc: Be and Si doping studies in $Al_xGa_{1-x}As$ and new results on persistent photoconductivity, J.Vac.Sci.Technol. B3 (1985) 2, 644-648

|69| Briones,F., D.Golmayo, L.Gonzalez, and J.L.de Miguel: SiH_4 doping of MBE GaAs and $Al_xGa_{1-x}As$, J.Vac.Sci.Technol. B3 (1985) 2,568-571

|70| Tandon,J.L., and Y.C.M.Yeh: Large-scale growth of GaAs epitaxial layers by MO CVD, J.Electrochem.Soc. 132 (1985) 3, 662-668

|71| Razuvaev,G.A., B.G.Gribov, G.A.Domrachev, and B.A.Salamalin: Metalorganicheskye soedinenya v elektronike, Nauka, Moscow 1977.

|72| Manasevit,H.M.: Single-crystal GaAs on insulating substrates, Appl.Phys.Lett. 12 (1968) 4, 156-159

|73| Coates,G.E., M.L.H.Green, and K.Wade: Organometallic compounds - the main group elements, v.1, 3rd ed., Methuen, London 1967.

|74| Stringfellow,G.B.: A critical appraisal of growth mechanism in MO VPE, J.Crystal Growth 68 (1984) 1, 111-122

|75| Stringfellow,G.B.: Epitaxy,Rep.Progr.Phys. 45 (1982) 2, 469-525

|76| Thrush,E.J., J.E.A.Whiteaway, G.Wale-Evans, D.R.Wight, and A.G.Cullis: Compositional transients in MO CVD grown III-V heterostructures, J.Crystal Growth 68 (1984) 1, 412-421

|77| Leys,M.R., C.van der Opdorp, M.P.A.Viegers, and H.J.Talen-van der Mheen: Growth of multiple thin layer structures in the GaAs-AlAs system using a novel VPE reactor, J.Crystal Growth 68 (1984) 1, 431-436

|78| Theeten,J.B., F.Hottier, and J.Hallais: Ellipsometric assessment of GaAlAs/GaAs epitaxial layers during their growth in an OM VPE system, J.Crystal Growth 46 (1979) 1, 245-252

|79| Hottier,F., J.Hallais, and F.Simondet: In situ monitoring by ellipsometry of metalorganic epitaxy of AlGaAs/GaAs superlattice, J.Appl.Phys. 51 (1980) 3, 1599-1602

|80| Chu,W.K., C.K.Pan, and C.A.Chang: Superlattice interface and lattice strain measurement by ion channeling, Phys.Rev.B 28 (1983) 7, 4033-4036

|81| Chang,C.A., and W.K.Chu: Channeling studies of Ge-GaAs superlattices grown by MBE, Appl.Phys.Lett. 42 (1983) 5, 463-465

|82| Petroff,P.M., R.C.Miller, A.C.Gossard, and W.Wiegmann: Impurity trapping, interface structure, and luminescence of GaAs quantum wells grown by MBE, Appl.Phys.Lett. 44 (1984) 2, 217-219

|83| Lewis,B.F., T.C.Lee, F.J.Grunthaner, A.Madhukar, R.Fernandez, and J.Maserjian: RHEED oscillation studies of MBE growth kinetics and lattice mismatch strain-induced effects during InGaAs growth on GaAs(100), J.Vac.Sci.Technol. B2 (1984) 3, 419-424

|84| Tsang,W.T., T.H.Chiu, S.N.G.Chu, and J.A.Ditzenberger: $GaSb_{0.5}As_{0.5}/Al_{0.35}Ga_{0.65}Sb_{0.48}As_{0.52}$ superlattice matched to InP prepared by MBE, Appl.Phys.Lett. 46 (1985) 7, 659-661

|85| Rao,E.V.K., H.Thibierge, F.Brillouet, F.Alexandre, and R. Azoulay: Disordering of $Ga_{1-x}Al_xAs$-GaAs quantum well structures by donor sulfur diffusion, Appl.Phys.Lett. 46 (1985) 9, 867-869

|86| Kawabe,M., N.Shimizu, F.Hasegawa, and Y.Nannichi: Effects of Be and Si on disordering of the AlAs/GaAs superlattice, Appl. Phys.Lett. 46 (1985) 9, 849-850

|87| Herman,M.A., and M.Pessa: $Hg_{1-x}Cd_xTe$-$Hg_{1-y}Cd_yTe$ ($0<x,y<1$) heterostructures-properties, epitaxy, and applications, J. Appl.Phys. 57 (1985) 8, 2671-2694

|88| Chow,P.P., and D.Johnson: Growth and characterization of MBE grown HgTe-CdTe superlattices, J.Vac.Sci.Technol. A3 (1985) 1, 67-70

|89| Furdyna,J.K.: Diluted magnetic semiconductors-an interface of semiconductor physics and magnetism, J.Appl.Phys. 53 (1982) 11, 7637-7643

|90| Pessa,M., and O.Jylhä: Growth of $Cd_{1-x}Mn_xTe$ films with $0<x<0.9$ by atomic layer epitaxy, Appl.Phys.Lett. 45 (1984) 6 646-648

|91| Bicknell,R.N., R.W.Yanka, N.C.Giles-Taylor, D.K.Blanks, E.L. Buckland, and J.F.Schetzina: $Cd_{1-x}Mn_xTe$-CdTe multilayers grown by MBE, Appl.Phys.Lett. 45 (1984) 1, 92-94

|92| Kolodziejski,L.A., T.C.Bonsett, R.L.Gunshor, S.Datta, R.B. Bylsma, W.M.Becker, and N.Otsuka: MBE of diluted magnetic semiconductor $Cd_{1-x}Mn_xTe$ superlattices, Appl.Phys.Lett. 45 (1984) 4, 440-442

|93| Bicknell,R.N., N.C.Giles-Taylor, D.K.Blanks, R.W.Yanka, E.L. Buckland, and J.F.Schetzina: Properties of $Cd_{1-x}Mn_xTe$-CdTe superlattices grown by MBE, J.Vac.Sci.Technol. B3 (1985) 2, 709-713

|94| Kołodziejski,L.A., R.L.Gunshor, S.Datta, T.C.Bonsett, M. Yamanishi, R.Frohne, T.Sakamoto, R.L.Bylsma, W.M.Becker, and N.Otsuka: MBE growth of films and superlattices of diluted magnetic semiconductors, J.Vac.Sci.Technol. B3 (1985) 2, 714-717

|95| Datta,S., J.K.Furdyna, and R.L.Gunshor: Diluted magnetic semiconductor superlattices and heterostructures,Superlattices and Microstructures, 1 (1985) 4, 327-334

|96| Zhang,X.C., S.K.Chang, A.V.Nurmikko, L.A.Kolodziejski, R.L. Gunshor, and S.Datta: Interface localization of excitons in CdTe/$Cd_{1-x}Mn_xTe$ multiple quantum wells, Phys.Rev.B 31 (1985) 6, 4056-4059

|97| Kinoshita,H., and H.Fujiyasu: PbTe-$Pb_{1-x}Sn_xTe$ superlattices prepared by a hot wall technique, J.Appl.Phys. 51 (1980) 11, 5845-5846

|98| Ambrosch,K.E., H.Clemens, E.J.Fantner, G.Bauer, M.Kriechbaum, P.Kocevar, and R.J.Nicholas: Structural and electronic properties of PbTe/$Pb_{1-x}Sn_xTe$ superlattices, Surface Sci. 142 (1984) 2, 571-578

|99| Ploog,K., and G.H.Döhler: Compositional and doping superlattices in III-V semiconductors, Advances in Phys. 32 (1983) 3, 285-359

|100| Döhler,G.H., and P.Ruden: Properties of nipi doping superlattices in III-V and IV-VI semiconductors, Surface Sci. 142 (1984) 1, 474-485

|101| Ploog,K., A.Fischer, and H.Künzel: The use of Si and Be impurities for novel periodic doping structures in GaAs grown by MBE, J.Electrochem.Soc. 128 (1981) 2, 400-410

|102| Jung,H., and K.Ploog: Influence of material design parameters on radiative recombination in GaAs doping superlattices grown by MBE, Appl.Phys. A37 (1985) 1, 83-87

|103| Kinoshita,H., T.Sakashita, and H.Fujiyasu: (p/n)PbTe multiple layer films prepared by a hot wall technique, J.Appl. Phys. 52 (1981) 4, 2869-2871

|104| Črobar,R.: Diffusion in semiconductor superlattices-decay of dopant concentration profiles, Czechoslovak J.Phys. B33 (1983) 1, 40-48

|105| Bean,J.C., T.T.Sheng, L.C.Feldman, A.T.Fiory, and R.T. Lynch: Pseudomorphic growth of Ga_xSi_{1-x} on silicon by MBE, Appl.Phys.Lett. 44 (1984) 1, 102-104

|106| Fiory,A.T., J.C.Bean, L.C.Feldman, and I.K.Robinson: Commensurate and incommensurate structures in MBE grown Ge_xSi_{1-x} films on Si(100), J.Appl.Phys. 56 (1984) 4, 1227-1229

|107| Fiory,A.T., J.C.Bean, R.Hull, and S.Nakahara: Thermal relaxation of metastable strained-layer Ge_xSi_{1-x}/Si epitaxy, Phys.Rev. B 31 (1985) 6, 4063-4065

|108| Gibson,J.M., R.Hull, J.C.Bean, and M.M.J.Treacy: Elastic relaxation in transmission electron microscopy of strained-layer superlattices, Appl.Phys.Lett. 46 (1985) 7, 649-651

|109| Krishnamurthy,S., and J.A.Moriarty: Electronic structure of silicon superlattices, Superlattices and Microstructures 1 (1985) 3, 209-215

|110| Abeles,B., and T.Tiedje: Amorphous semiconductor superlattices, Phys.Rev.Lett. 51 (1983) 21, 2003-2006

|111| Abeles,B., T.Tiedje, H.C.Stasiewski, H.W.Deckman, P.D.Persans, K.S.Liang, and C.B.Roxlo: Amorphous semiconductor superlattices, Superlattices and Microstructures 1 (1985) 2, 115-118

Chapter 3. THEORY OF SEMICONDUCTOR SUPERLATTICES

Along with the growing interest in semiconductor superlattices, various theoretical schemes have been proposed to study the nature of the electronic states within these structures.

The difference between the electronic properties of a superlattice and a bulk semiconductor arises from the periodic potential which is associated with the superstructure. This can be easily observed in a type I compositional superlattice, where the periodic modulation of the band-edges of the constituent semiconductors deeply modifies the positions of the allowed energy levels for the charge carriers in the superlattice compared with their positions in the bulk semiconductors. In particular, the continuum of the allowed electronic states of the host bulk material with the narrower energy gap become split into one-dimensional minibands separated by finite band gaps. Reciprocally, superlattice states may be allowed for energies corresponding to evanescent states in the host bulk material with the wider energy gap.

The superlattice wave functions are usually constructed as linear combinations of Brillouin-zone centre bulk states with slowly varying envelope modifications reflecting the superstructure periodicity along the superlattice axis. The properties of these superlattice functions are of major relevance when evaluating the strength and the shape of the optical absorption and radiative recombination spectra. Thus, the super-periodicity changes also considerably the optical properties of the superlattices in comparison to the bulk constituent materials.

There are many different theoretical problems important from the point of view of a complete physical description of the large variety of superlattice structures fabricated at present. A precise presentation of all of the physical models and calculation techniques used in the theory of superlattices goes far beyond the scope of this book. Therefore, only the most fundamental theoretical problems will be presented in detail, while the more advanced problems will be mentioned only superficially.

3.1. A general survey of the theory relevant to semiconductor superlattices

Three main groups of theoretical problems related to semiconductor superlattices can be distinguished at present. The first group comprises the electronic properties of superlattice structures and is mainly related to calculations of the energy dispersion relations along the superlattice growth axis. The second group includes optical properties, i.e., photoabsorption, and radiative recombination (luminescence). It is concerned with calculation and with analysis of the optical matrix elements describing quantum transitions of the superlattice between the energy states involved in the optical phenomenon analysed. The final group of theoretical problems deals with the so-called collective excitations in semiconductor superlattices.

3.1.1. Electronic properties

From a basic theoretical point of view it would be desirable to calculate the electronic properties of a superlattice directly by solving the corresponding band structure problem for the three-dimensional crystal lattice. There is no principal difficulty for traditional band structure techniques |1|,|2| to treat the superlattices. In the case of a compositional superlattice, the two different materials comprised in the heterostructure have usually matched lattice constants |3|, or if their lattices are not matched the layers are strained. Anyhow, the bulk periodicity persists parallel to the layers. The new periodicity in the growth direction, however, increases the unit cell, relevant to the band structure calculations, considerably. The size of the Hamiltonian matrix becomes, therefore, very large inducing strong limitations on the maximum number of layers in the periodicity interval that can be treated by traditional three-dimensional methods |4|.

Ihm et al.|5| have reported superlattice bands for a 12 layer GaSb-12 layer InAs heterostructure which were calculated using the self-consistent pseudopotential method. Also traditional tight-binding calculations |6-8|, yield very large matrices for typical experimentally interesting slab thicknesses. The matrices dimensions are equal to the number of orbitals per atom times the number of atoms in the unit cell. The presence of the interfaces destroys translational symmetry in the direction of the superlattice axis, and thus the unit cell size becomes infinite.

In order to circumvent this difficulty new methods were developed, which correctly treat the semi-infinite nature of the problem |9|, |10|, yet allow to solve for the electronic structure of such Hamiltonians much more efficiently without having to refer to large matrices. This was done in a variety of ways, including the use of bulk and surface Green's functions, and/or transfer matrices.

The transfer matrix approach has been applied, e.g., in the study of the type II InAs-GaSb superlattice |11| or in the reduced Hamiltonian method proposed for a convenient treatment of: (i) two-dimensional electronic systems at heterojunction interfaces, (ii) quantum wells, and (iii) different types of compositional superlattices |12-15|. The resolvent method based on a Green's function formulation of scattering theory suitable for treating locally perturbed semiconductors |9| has been applied to the type I GaAs-$Al_xGa_{1-x}As$ superlattice |4|.

Taking into consideration the fact that usually the characteristic length d describing the artificial crystal periodicity in superlattices is much larger than the crystal lattice constant a_o which represents the natural crystal periodicity, i.e., that $a_o \ll d$, the superlattice electronic states may also be described within the framework of the envelope-function approximation |16-20|. The tight-binding approach focuses the attention on atomic-like properties (i.e., at the a_o scale) and builds the superlattice states from host atomic site to host atomic site. The envelope-function scheme takes the natural periodicity a_o into account by means of an effective-mass approximation, whereas the superperiodicity d acts upon the envelope functions which are slowly varying at the scale of a_o.

The tight-binding calculations are exact in essence but to be accurate they still require quite elaborate computational efforts, so they are ill suited for actual layer thicknesses (i.e., for $d \gg a_o$). In the envelope-function scheme, some microscopic (i.e., at the scale of a_o) information is lost. In particular, the details of interfaces between two consecutive layers and the exact chemical nature of actual host layers (e.g., which are the chemical elements of the two terminating planes of a given layer) are left undefined. In this coarse-grained description, which is reasonable only for sufficiently thick host layers (d $\approx$ 10 nm) |21| each layer is an effective medium of which only some gaps and an interband-matrix element are known. The electron dynamics in the superlattice is described by the envelope functions which are eigenstates of an effective (eventually multiband) Hamiltonian. The microscopic information in this scheme is embodied into the Γ energy gaps, Γ spin-orbit coupling and Kane matrix element, which are known from experiments on bulk materials.

Despite its apparent crudeness, the envelope-function approximation is in excellent agreement with experiments, for both the type I and the type II superlattice systems |21|. This is not very surprising as long as the experiments have been performed on crystals with d >> a_0 for which the interface planes represent a relatively minor detail compared with the "bulk" of each host layer. It should be emphasized that the envelope-function approximation method is adequate for a superlattice state whose energy is in close proximity to a band-edge of one of the bulk materials and if only one or two bulk states in each material are expected to dominate |15|. Nevertheless, this method is very useful in practice because it is simple enough so that the relationship between the superlattice and bulk wave functions is easily extracted, yet it leads to results that are well confirmed by many experiments.

In contrast to the theoretical treatment of compositional superlattices, a particular advantage of doping superlattices, with respect to the theoretical description of their electronic structure, is that one is dealing in this case with a homogeneous semiconductor which is only modulated by the superposition of a periodic superlattice potential. For most cases of interest, the effective-mass approximation provides a very satisfactory description. The space-charge potential varies slowly over distances of the order of the bulk lattice constant and the considered energies of the bands are sufficiently close to the band-edges. Therefore, the neglect of corrections due to nonparabolicity of the bands is justified |3|.

Another simplifying approximation, which is often used in the theory of doping superlattices, is to replace the space-charge distribution of the individual doping atoms by homogeneous distribution functions, $eN_D(z)$ for donors, and $-eN_A(z)$ for acceptors, varying periodically in superlattice direction z. In doing so, one neglects the spatial potential fluctuations which result from the random distribution of impurities in the doped regions, and the point charge character of the impurities which may lead to bound impurity states or to the formation of impurity bands.

The significance of the deviations, which are actually met in doping superlattices, from the simplified homogeneous space-charge model has been analyzed in Ref. 22 in the framework of the local-density-functional formalism |23|,|24|. It has been shown that the random potential fluctuations are resonably well screened by free carriers and that impurity band effects are usually negligible for the electrons. For holes, in contrast, it has been found that they normally populate an acceptor impurity band rather than valence subbands. The quite different behaviour originates from the large difference between the conduction-electron mass and the heavy-hole mass. This leads to a small donor binding energy, large subband distances and Fermi energies in the former case, but a comparatively large acceptor binding energy and small heavy-hole subband distances in the latter one.

The presented argument applies mainly to typical direct-gap III-V semiconductors, like GaAs. In other materials, e.g., in IV-VI compounds the situation is different because of the special band structure with a direct gap in momentum space at the L-point of the Brillouin zone, and the absence of bound shallow impurity states |3|,|25|.

A further simplification of the theoretical framework required for describing the electronic properties of superlattice structures seems to be possible for amorphous superlattices, made of hydrogenated amorphous semiconductors |26|.

One of the most intriguing questions regarding the electronic properties of these rather exotic superlattice materials is whether they exhibit quantum size effects similar to those observed in crystalline superlattices. Evidence for the existence of these effects in a-Si:H layers less than 4 nm thick, sandwiched between barrier layers of a-SiN_x:H has been gained from optical and electrical measurements performed with the a-Si:H/a-SiN_x:H superlattice |27|. In order to describe the observed effects theoretically in a way consistent with the experimental data, a simple energy band model in the effective-mass approximation for free electrons and holes moving in a one-dimensional periodic potential, similar to that which is characteristic for the type I compositional superlattice (see, Fig.1.1.a) has been proposed |27|,|28|. According to this model, confinement of a free carrier exhibiting an effective mass m^* in a one-dimensional quantum well, with depth U and width L, of the amorphous superlattice increases the ground-state energy of the carrier by a value E_i which depends on L (E_i increases when L decreases) for well layer thicknesses less than 4 nm, and is independent of L, for well thicknesses exceeding 4 nm |27|. This means that the coherence length of the carrier wave functions in the superlattice wells is of the order of 4 nm, and that the quantum size effect in amorphous superlattices can be observed only for well layer thicknesses less than 4 nm.

In the case of the a-Si:H/a-SiN_x:H superlattice excellent agreement between the theoretical predictions and the experimental data has been obtained when the following assumptions have been made:

(i) $U = 0.5\ \Delta E_g = 1.05$ eV for both, the conduction and the valence band-edge discontinuities

(ii) $m_e^* = m_h^* = 1$ (the effective mass is equal to the mass of a free electron). These parameters of the model are, unfortunately, not unique, since the experimental data could have been fitted by a different combination of band-edge discontinuities and the effective masses.

The same model was also applied with success to other types of amorphous silicon-based superlattices, where the difference in the band gaps is smaller than in the a-Si:H/a-SiN_x:H system, and where tunneling of carriers between wells is important.

So far, no more advanced theory relevant to the silicon-based amorphous superlattices has been published.

3.1.2. Optical properties

The optical properties of compositional superlattices, have so far been analysed in two approximations. In the first, the superlattice states are described in the envelope-function formalism |21| whereas the tight-binding approximation framework is applied in the second |29|. The most important quantities that have to be calculated first, when analysing the mentioned optical properties, are the momentum matrix elements relevant to the optical transitions between superlattice eigenstates associated with the n-th (conduction band) and m-th (valence band) subbands. We can restrict our considerations only to the absorption phenomenon without losing generality, because the matrix elements related to the absorption may also be used to analyse other optical properties.

For transitions between given valence and conduction subband states $|m,\vec{k}_t,q\rangle$ and $|n,\vec{k}_t,q\rangle$, the absorption coefficient of a superlattice is given by (apart from a constant factor) |29|,|30|

$$\alpha(\hbar\omega) = \omega^{-1} \sum_{k_t,q} \sum_{n,m} |\vec{e}\, P_{mn}(\vec{k}_t,q)|^2 . \delta[E_n(\vec{k}_t,q) - E_m(\vec{k}_t,q) - \hbar\omega] \tag{3.1.1}$$

where $\vec{e}$ denotes the direction vector of light polarization and $P_{mn} \equiv \langle n,\vec{k}_t,q|\vec{p}|m,\vec{k}_t,q\rangle$ denotes the momentum matrix element between superlattice eigenstates associated with the n-th and the m-th subbands. The symbols $\vec{k}_t$ and q designate the projections of the wave vector in the plane parallel to the interface and along the axis perpendicular to the interface (the z-axis), respectively, whereas $E_n(\vec{k}_t,q)$ stands for the energy associated with the eigenstate $|n.\vec{k}_t,q\rangle$. To evaluate $\alpha(\hbar\omega)$ for a realistic superlattice, one needs to know the detailed energy spectra and the momentum matrix elements between various electronic states.

In the envelope-function approximation |21| the following basic assumptions are made:

(i) only perfectly lattice-matched III-V and II-VI compound direct-gap superlattice constituent materials are considered,

(ii) only superlattice states which are close enough to the host materials Γ points take part in the optical transitions, so only the usual Γ_6, Γ_7 and Γ_8 host-materials band-edges are considered as contributing significantly to the superlattice wave function,

(iii) the Γ-point periodic parts of the hosts Bloch functions u_v $(1 < v < 8)$ are identical in all layers of the superlattice,

(iv) any inversion-asymmetry effect associated with the zinc-blende lattice is neglected,

(v) the superlattice wave functions are written as

$$\psi(\vec{r}) = \sum_{v=1}^{v=8} u_v(\vec{r}) F_v(\vec{r}) \tag{3.1.2}$$

where the envelope functions $F_v(\vec{r})$, slowly varying at the scale of a_o, are the solutions of an 8 x 8 effective Hamiltonian obtained by replacing k_z in the bulk quasi-Ge $\vec{k}\cdot\vec{p}$ Hamiltonian by the operator $-i\hbar\,\partial/\partial z$ (of the momentum z-component) for the kinetic energy and adding a superlattice potential energy term $V_{vv}(z)$ which is diagonal in v and represents the jump of the v-th band-edge when going from the one superlattice constituent material to the other one if flat band conditions prevail. Any envelope function $F_v(z)$, $1<v<8$, is the solution of a 1 x 1 effective Hamiltonian H_v obtained by projecting the 8 x 8 effective Hamiltonian onto the v-th edge. It should be noticed that the Hamiltonian H_v contains only even powers of $\partial/\partial z$ |17|,|18|. It is also obvious that any microscopic details, as those occurring at the interfaces, are washed out in the envelope-function scheme.

Let us consider two consecutive A,B layers of a compositional superlattice (see,Fig.3.1.) with O_A, O_B being the centres of the layers and P_A, P_B being the axis bisecting these layers. The wave functions (3.1.2) used for the shown type I and type II superlattice structures can be expressed, in order to calculate the interband matrix elements, in the following more explicite form |21|

$$\psi^{e,h}_{nk_tq}(\vec{r}_t,z) = \sum_v S^{-1/2}\, C^{e,h}_{vnk_tq}\, U_v(\vec{r}_t,z)\exp(i\vec{k}_t\vec{r}_t) f^{e,h}_{vnk_tq}(z) \qquad (3.1.3)$$

e and h refer here to conduction and valence states, respectively, n is a subband index, $\vec{r}_t$ is a two-dimensional vector in the layer planes of area S,C_{vnk_tq} is a normalization coefficient, U_v is the quickly varying periodic part of the v-th host material Bloch function, and f_{vnk_tq} is the associated z-dependent part of the envelope function F_v (see,Fig.3.1).

In most cases of actual interest, the superlattice subbands widths are smaller than the energy separation between the various levels of the isolated quantum wells from which they originate (no band-mixing effects are taken into consideration). Therefore, a tight-binding description of the envelope functions $f_{vnk_tq}(z)$ occurring in Eq. (3.1.3) may be used. As a result these functions are written in the form

$$f^e_{nq}(z) = N^{-1/2} \sum_{p=-\infty}^{p=\infty} \phi^e_n(z-pd)\exp(iqpd) \qquad (3.1.4)$$

$$f_{mq}^{h}(z) = N^{-1/2} \sum_{p=-\infty}^{p=\infty} \phi_n^h(z-pd-\sigma d/2)\exp(iqpd) \tag{3.1.5}$$

where N is the number of superlattice unit cells in the crystal, and $\sigma = 0(1)$ in type I(II) superlattices. $\phi_n^e(z-pd)$ and $\phi_m^h(z-pd-\sigma d/2)$ are here the eigenfunctions of the bound states of the conduction - and valence - isolated quantum wells at the crystal sites labeled p, which exhibit a very weak dependence on the wave vector $\vec{k}_t$ (usually ignored in the envelope-function approximation). The overlap integral $\langle f_{nq}^e | f_{mq}^h \rangle$ of these functions, which determines the optical absorption in the superlattice is equal to

$$Q_{nm}(q) = \sum_{p=-\infty}^{p=\infty} \exp(iqpd) \cdot \mu_{nm}(p) \tag{3.1.6}$$

where

$$\mu_{nm}(p) = \int_{-\infty}^{\infty} [\phi_n^e(z)\phi_m^h(z-pd-\sigma d/2)]dz \tag{3.1.7}$$

For type I superlattices $\sigma=0$, and with the use of the symmetry properties of ϕ's in these superlattices, Eq. (3.1.6) can be rewritten as

$$Q_{nm}(q) = \mu_{nm}(0) + \begin{cases} 2\sum_{p=1}^{p=\infty} \cos(qpd)\cdot\mu_{nm}(p), & n+m \text{ is even,} \\ 2i\sum_{p=1}^{p=\infty} \sin(qpd)\cdot\mu_{nm}(p), & n+m \text{ is odd,} \end{cases} \tag{3.1.8}$$

$\mu_{nm}(0)$ is nonvanishing only if m+n is even, the diagonal terms m = n being much more intense than the off-diagonal ones. For $p \neq 0$ $\mu_{nm}(p)$ has no selection rule but is exponentially small as compared to $\mu_{nm}(0)$. To a very good approximation $Q_{nm}(q)$ is independent of q and equal to $\mu_{nm}(0)$. It can be concluded from Eq. (3.1.8), that:

(i) if at q = 0 the optical transition is parity allowed (m+n is even), it remains allowed at $q = \pi/d$, and

(ii) if it is forbidden at q=0 (m+n is odd), it remains forbidden at $q = \pi/d$.

For type II superlattices $\sigma = 1$, so Eq. (3.1.6) may be rewritten in the form

$$Q_{nm}(q) = \sum_{p=0}^{p=\infty} \exp(iqpd)\{1+(-1)^{n+m}\exp[-i(2p+1)qd]\}\mu_{nm}(p) \tag{3.1.9}$$

None of the μ's vanishes and all are exponentially small, being nonzero only through the leakage of ϕ's functions outside the wells where they are centered. The general parity properties of the optical transition are now different from the case of type I superlattices.

(i) If the transition is parity allowed at q=0 (m+n is even) then it becomes parity forbidden at $q = \pi/d$, and

(ii) if m+n is odd, the transition is allowed at $q = \pi/d$ and forbidden at q=0. $Q_{nm}(q)$ depends now strongly on q. To the lowest order, one has

$$Q_{nm}(q) = \begin{cases} 2\,\mu_{nm}(0)\exp(-iqd/2)\cos(qd/2), & m+n \text{ is even} \\ 2i\,\mu_{nm}(0)\exp(-iqd/2)\sin(qd/2), & m+n \text{ is odd} \end{cases} \qquad (3.1.10)$$

Using the expressions (3.1.8) and (3.1.10) for the overlap integrals, one can calculate the absorption, as well as luminescence, line shapes in type I and type II superlattices in the envelope-function approximation |21|,|31|.

It should be amphasized that the parity selection rules, presented above in the envelope-function approximation for the type I and type II superlattices, exist only because the noncentrosymmetric nature of the III-V or II-VI unit cells, as well as the effects associated with the layer edges, have been neglected. These two simplifying assumptions may seem severe at first glance. However, the quasi-Ge model used in this approximation is known to work very well in bulk III-V and II-VI materials. As far as edge effects are concerned, they should be small for thick ($\approx$10 nm), practically the most common, superlattice layers. In fact the only transitions seen in absorption experiments as intense lines, have been those which fulfill the parity selection rules predicted by this theory |31|,|32|.

In the second theoretical approximation used to study the optical interband transitions in type I and type II compositional superlattices a realistic tight-binding model is used |15|,|29|. According to this model the electronic eigenstates of a superlattice are expanded in terms of planar atomic orbitals $\psi_\alpha(\vec{k}_t,l)$. These are two-dimensional Bloch sums of atomic orbitals in a plane parallel to the interface, whereas α denotes the orbital symmetry (s,x,y,z, ...), and l denotes the position of the atomic plane. The expansion is written in the form

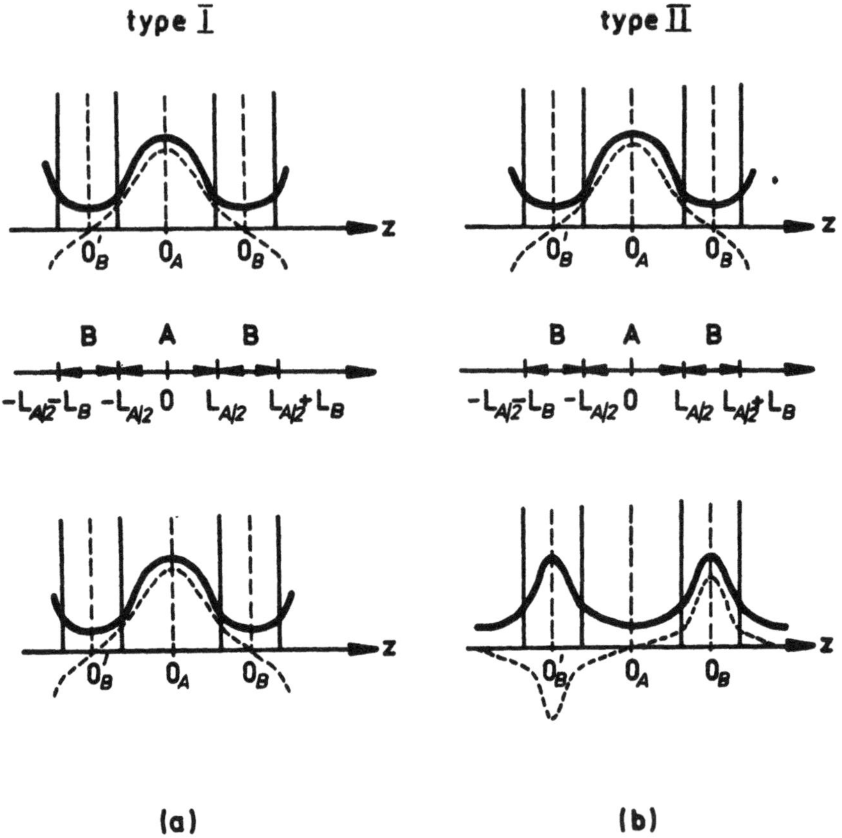

Fig.3.1. Schematic illustration of the envelope functions for conduction and valence ground subbands at q=0 (solid lines) and q = π/d (dashed lines) in type I (a) and type II (b) superlattices (after Voisin et al.,Ref.21)

$$|n,\vec{k}_t,q\rangle = \sum_{\alpha,l} A^n_\alpha(k_t,q,l)\,|\psi_\alpha(\vec{k}_t,l)\rangle \qquad (3.1.11)$$

where l runs over all atomic planes of the whole superlattice. The energy spectra $E_n(\vec{k}_t,q)$ and the expansion coefficients $A^n_\alpha(\vec{k}_t,q,l)$ (also known as the envelope functions) are usually calculated in this model by the "reduced Hamiltonian" method as described in |14|.

Five atomic orbitals (labeled with s^*, s, x, y, z) per atomic site are used. These orbitals are sufficiently localized so that the momentum matrix elements occurring in the Eq. (3.1.1) can be neglected when they are related to atoms separated by more than first-neighbour distance. In general these momentum matrix elements can be written in terms of the coefficients $A_{\alpha}^{m}(\vec{k}_t,q,l)$ and the momentum matrix elements between planar atomic orbitals

$$P_{mn}(k_t,q) = \sum_{l,\alpha,\alpha' l'} A_{\alpha}^{m^*}(\vec{k}_t,q,l)A_{\alpha'}^{n}(\vec{k}_t,q,l') \times \times \langle\psi_{\alpha}(\vec{k}_t,l)|\vec{p}|\psi_{\alpha'}(\vec{k}_t,l')\rangle \qquad (3.1.12)$$

Furthermore, according to the model, the low-lying conduction subband states are almost purely s-like whereas the valence subband states are almost purely p-like. Therefore, the optical matrix elements involving transitions between two s-like orbitals or two p-like orbitals can be ignored. Also the transitions involving the s^*-like orbitals are ignored. Consequently, only four independent parameters remain to be determined for each constituent material of the considered superlattice. These parameters are

$$\begin{aligned} P_{aa} &= i\langle\phi_s(\tau_a)|p_x|\phi_x(\tau_a)\rangle(2/m_e)^{1/2} \\ P_{cc} &= i\langle\psi_s(\tau_s)|p_x|\phi_x(\tau_c)\rangle(2/m_e)^{1/2} \\ P_{ac} &= i\langle\phi_s(\tau_a)|p_x|\phi_x(\tau_c)\rangle(2/m_e)^{1/2} \\ P_{ca} &= i\langle\phi_s(\tau_c)|p_x|\phi_x(\tau_a)\rangle(2/m_e)^{1/2} \end{aligned} \qquad (3.1.13)$$

where $\phi_{\alpha}(\tau)$ denotes an atomic orbital of symmetry α, located at the position τ, τ_a and τ_c indicate the anion and the cation positions in the unit cell, and m_e is the free-electron mass. The other momentum matrix elements can be infered by symmetry.

The matrix elements $\langle\psi_{\alpha}(\vec{k}_t,l)|\vec{p}|\psi_{\alpha'}(\vec{k}_t,l')\rangle$ can be written in terms of the four parameters given by Eqs. (3.1.13). These parameters can be directly related to the optical matrix elements associated with various interband transitions for the constituent bulk materials, which can be either measured experimentally or calculated by other reliable theoretical techniques suitable for being used for analyzing the optical properties of bulk semiconductors |33|. The relations are provided by writing the bulk Bloch states of various bands at $\vec{k}=0$ in terms of the tight-binding orbitals used in this model, and calculating the interband transitions with these states |29|.

Applying this theoretical procedure, the optical matrix elements of transitions from several valence subbands to several conduction subbands could be calculated as functions of the well width and of the wave vector for GaAs-$Al_xGa_{1-x}As$ and InAs-GaSb superlattices. It has been found that the mixing of the heavy- and light--hole components in the superlattice states gives rise to a large variation in the optical matrix elements as $\vec{k}$ moves away from the zone centre. The band mixing in conjunction with the exciton effect leads to weak structures in the absorption spectrum which can account for the "forbidden" transitions observed in several recent experiments |34|,|35|. These results differ considerably from the predictions of the envelope-function approximation, where band mixing effects are excluded from the considerations.

The absorption of photons in doping superlattices is in principle possible once the photon energy $\hbar\omega$ exceeds the effective energy gap E_g^{eff}. In practice, however, the interband matrix elements for transitions between the uppermost valence and the lowest conduction subbands are extremely small for the case of large superlattice period and moderate doping concentrations |3|. The matrix elements, however, increase exponentially as $\hbar\omega$ approaches the gap of the unmodulated bulk semiconductor E_g. It is also clear that the absorption coefficient of a doping superlattice as a function of the photon energy will be a function of the effectiv energy gap and, therefore, of the quasi-Fermi-level difference $\Delta E_{F_{n,p}}$ since the optical matrix elements will be strongly influenced by variations of the periodic space-charge potential |3|.

As a first approximation to the problem of optical absorption in doping superlattices, the continuum approach may be used. In this approach the absorption coefficient is given by the bulk value α^b, averaged over the internal local field distribution in the superlattice

$$\alpha(\omega, \Delta E_{F_{n,p}}) \approx d^{-1} \int_0^d \alpha^b \, dz \tag{3.1.14}$$

This approach is valid for superlattices with long period and a moderate doping level, because the spatial variation of the internal electric fields is there rather slow and the envelope wave functions of the true subbands do not differ significantly from the Airy functions obtained for homogeneous fields near the band edges.

In a more exact approach |3| the absorption coefficient of a doping superlattice is related to the momentum matrix element between different subbands in the valence and in the conduction bands of the host crystal

$$\alpha(\omega) = (\omega/nc)2(4\pi e^2/m_o^2\omega^2 d) \sum_{v=lh,hh} \sum_{\mu,\nu} |\langle c,\mu|\vec{p}|v,\nu\rangle|^2 N^{(2)}_{v\nu,c\mu}(\hbar\omega) \tag{3.1.15}$$

where n is the refractive index of the host crystal, c the velocity of light, and $N_{n\nu,c\mu}(\hbar\omega)$ is the joint density of states per superlattice period. At photon energies near the threshold, the absorption coefficient is dominated by transitions from the uppermost light hole to the lowest conduction subband if only one conduction subband is populated. For a doping superlattice with no intrinsic layer ($d_i=0$) and with symmetric doping ($d_n = d_p$ and $N_D = N_A$), the square of the effective interband matrix element is

$$|\langle c,0|p|lh,0\rangle|^2 \simeq |p^{bulk}_{lh,c}|^2 \exp[-(d/2)^2/(\alpha_c^2 + \alpha_{lh}^2)] \tag{3.1.16}$$

where $p^{bulk}_{lh,c}$ is the bulk interband matrix element of momentum at the Γ point, $\alpha_c = \hbar^{1/2}(m_c^*\omega_c)^{-1/2}$, $\alpha_{lh} = \hbar^{1/2}(m_{lh}^*\omega_{lh})^{-1/2}$, ω_c and ω_{lh} are the electron and light-hole bulk plasma frequencies. The corresponding expression for the heavy-hole contribution to the matrix element, with α_{lh} substituted by α_{hh} is negligibly small if the exponent in the Eq. (3.1.16) is considerably larger than unity. This results from the fact, that the more strongly localized heavy-hole subbands have an exponentially smaller overlap with the conduction subbands as compared to those of the light-hole wave functions.

With $\alpha_{lh} \approx \alpha_c$ one can rewrite Eq. (3.1.16) as

$$|\langle c,0|p|lh,0\rangle|^2 \approx |p^{bulk}_{lh,c}|^2 \exp[-2(E_g^o - E_g^{eff})/\hbar\omega_c] \tag{3.1.17}$$

From Eq. (3.1.15) it is obvious that $\alpha(\omega)$ increases stepwise whenever a new absorption process becomes energetically possible. The absorption coefficients of a GaAs doping superlattice, calculated as a function of photon energy for the ground state and two different excited states of this superlattice, are shown in Fig. 3.2 |3|. It is clear that the absorption coefficient changes as a function of the effective band gap E_g^{eff} ($\approx \Delta E_{F_{n,p}}$). The curves in Fig. 3.2. for $n^{(2)} = 6 \cdot 10^{11}$ cm^2 and $1.2 \cdot 10^{12}$ cm^{-2} corresponding to $\Delta E_{F_{n,p}}$ = 1.329 and 1.417 eV, respectively, exemplify this variation.

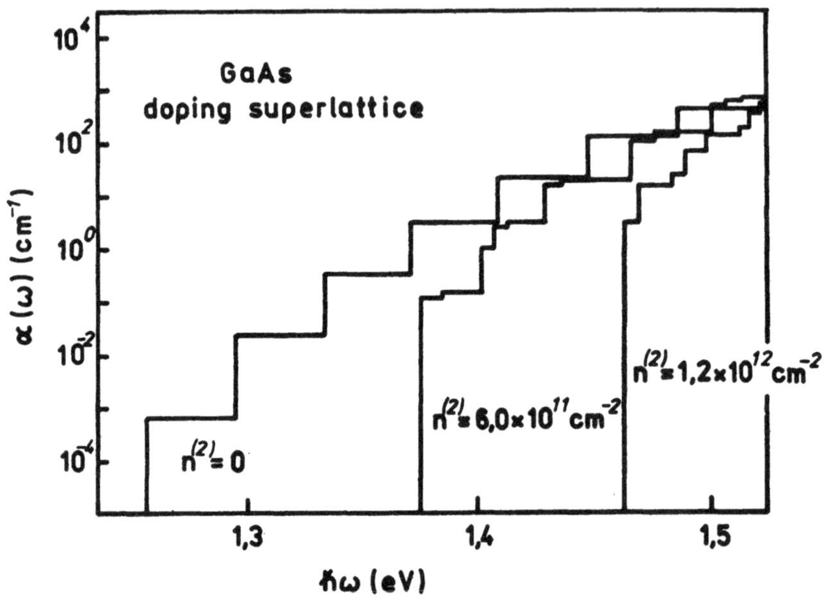

Fig.3.2. Absorption coefficient of a GaAs doping superlattice as a function of photon energy, calculated for the ground state ($n^{(2)} = p^{(2)} = 0$) and two different excited states ($n^{(2)} = p^{(2)} = 6.10^{11}$cm^{-2}) and ($n^{(2)} = p^{(2)} = 1.2 \cdot 10^{12}$ cm^{-2}). The design parameters are chosen such that $\hbar\omega_c = \hbar\omega_{lh}$ in the ground state. $N_D = (m_c^*/m_{lh}^*)N_A$, $d_n = (N_A/N_D)d_p$, $N_A = 1 \cdot 10^{18}$cm^{-3}, $d = d_n + d_p$ = 60 nm (after Ploog et al., Ref. 3).

A particularly interesting feature of the tunable absorption coefficient of the doping superlattice is its oscillatory behaviour as a function of $\Delta E_{F_{n,p}}$ at a fixed photon frequency. The origin of this theoretically predicted effect is the following |3|. The absolute values of the momentum matrix element in Eq. (3.1.15) increase with increasing $\Delta E_{F_{n,p}}$, since the overlap between the states $|c,\mu\rangle$ and $|v,\nu\rangle$ becomes larger due to the flattening of the self-consistent superlattice potential $V_{sc}(z)$. There is, however, a sudden drop of $\alpha(\omega,\Delta E_{F_{n,p}})$ whenever one of the contributions to the joint density of states $N_{v\nu,c\mu}(\hbar\omega)$ becomes energetically forbidden as a function of increasing $\Delta E_{F_{n,p}}$, which occurs, if for any value of μ and ν $(E_{c,\mu} - E_{v,\nu})_{\Delta E_{F_{n,p}}} = \hbar\omega$. This oscillatory behaviour is most pronounced if the subband spacing is large and if not too many momentum matrix elements contribute to the absorption coefficient. Thus, high doping concentrations and not too large period of the superlattice provide favourable conditions for observation of this effect. In doping superlattices with a large spatial period and moderate doping level, in contrast, the number of subbands contributing to the absorption may become so large that the step-like quantum structure is smeared out completely, even if only rather weak potential fluctuations are present. In this quasiclassical limit α can be calculated in the continuum approximation, as it has already been mentioned [see, Eq. (3.1.14)].

3.1.3. Collective excitations

So far, we have discussed the basic theory related to optical absorption in superlattices. Similar theoretical formalism can be used in analysing light emission from these multilayer structures. A different theoretical framework is, however, required when light scattering experiments have to be interpreted. These experiments have revealed a host of excitations in superlattices |36|, which can be understood only in terms of electronic collective modes.

The term "electronic collective modes of a superlattice structure" refers to the selfsustaining oscillations of the electron density in the structure. In order to obtain the response of the superlattice to an external perturbation, and hence the collective modes, one has to calculate the electron density response, or the current response. Most frequently this can be done by using the linear response theory |36|,|37|.

A superlattice, being a periodic system of two-dimensional electron gas (2DEG) layers |38| (compare the type I superlattice) or

2DEG and spatially separated two-dimensional hole gas (2DHG) layers |39| (compare the type II and doping superlattices) falls intermediately between two -, and three-dimensional systems. When the separation between the adjacent layers is very large, negligible coupling between them is expected, giving rise to two-dimensional behaviour. However, for vanishing separation between adjacent layers, the system is effectively three-dimensional. Thus, two special cases of particular interest exist, namely those of strong and weak coupling between adjacent layers. In the weak-coupling limit, the superlattice period is so large that the layers act independently of each other. In the strong coupling limit, on the other hand, the layers are so close together that the electronic collective excitation modes in one layer strongly affect those in the others.

Since the layer thickness of a superlattice can be controlled by MBE or MO CVD techniques, such a system provides a way of studying transitions from two-dimensional to three-dimensional behaviour in the collective mode spectrum as the layer thickness is reduced. The situation under an external magnetic field along the direction of superlattice growth (i.e., perpendicular to the 2DEG layers) is also interesting because the three-dimensional case of small separation should allow for extra modes that can propagate along the superlattice axis and the magnetic field directions, in addition to the familiar magnetoplasmon modes |37|. A comprehensive theoretical analysis of these problems has been presented in references |36-38| and |40-42|.

Using the linear response theory to calculate the density response of the superlattice system to an external perturbation, the expression for the dielectric tensor has been derived, the zeros of which yield the dispersion relations of the collective modes. A large variety of such modes, consisting of quasi-two-dimensional plasmons and magnetoplasmons, intersubband plasmons and phonon--plasmon modes, has been found.

Avoiding a too deep involvement into theoretical details we will present, as an example, only the theory of collective modes in the type I superlattice, in the absence of magnetic fields |36|.

The eigenstates of the superlattice are taken to be

$$|n,\vec{k}_t,l\rangle = \exp(i\vec{k}_t\vec{r}_t)F_n(z-ld) \qquad (3.1.18)$$

where n and l are the subband and the layer indices, respectively. The envelope function $F_n(z)$ satisfies the one-dimensional Schrödinger equation in the effective mass approximation

$$-(\hbar^2/2m_e^*)(d^2/dz^2)F_n(z) + V_{SL}^{eff}(z)F_n(z) = E_nF_n(z) \qquad (3.1.19)$$

Here $V_{SL}^{eff}(z)$ is the effective confining potential for a particular quantum well of the superlattice. It consists of the depletion layer potential (or in the case of intrinsic layer superlattices the band-edge discontinuity potential), which confines the electrons to the well, the Hartree potential, due to the averaged electrostatic potential of all the electrons in the well, and an exchange-correlation potential, assumed to be a local functional of the electron density. The Eq. (3.1.19) is therefore a Kohn-Sham equation |24|,|43|.

Let us now assume an external perturbing potential of the form

$$V^{ext}(\vec{r}_t,z,t) = V^{ext}(\vec{q},\omega,z)\exp(i\omega t - i\vec{q}\,\vec{r}_t) \qquad (3.1.20)$$

and use the Ehrenreich-Cohen self-consistent field prescription |44| to calculate the induced electron density

$$\delta n(\vec{r}_t,z,t) = \sum_{\alpha,\alpha'} \frac{f(E_{\alpha'}) - f(E_\alpha)}{E_{\alpha'} - E_\alpha - \hbar\omega} \langle\alpha|V|\alpha'\rangle \times$$

$$(3.1.21)$$

$$\times \langle\alpha'|\delta(\vec{r}_t - \vec{r}\,'_t)\delta(z-z')|\alpha\rangle$$

Here α is a composite index $\alpha = n,\vec{k}_t,l$, V is the sum of the external and induced potentials ($V = V^{ext} + V^{ind}$) and $f(E_\alpha)$ is a Fermi function. All induced quantities are taken to have the same form as V^{ext} in Eq. (3.1.20). Assuming no overlap between electrons in different layers, one can write

$$\langle\alpha|V|\alpha'\rangle = \delta_{l',l}\delta_{\vec{k}'_t,\vec{k}_t+\vec{q}}\langle n|V_l(\vec{q},\omega)|n'\rangle \qquad (3.1.22)$$

where

$$\langle n|V_l|n'\rangle = \int dz\, F_n(z-ld)V(\vec{q},\omega,z)F_{n'}(z - ld) \qquad (3.1.23)$$

From Eq. (3.1.20) one obtains

$$\delta n(\vec{q},\omega,z) = \sum_{nn'l} \pi_{nn'}(\vec{q},\omega)\langle n|V_l(\vec{q},\omega)|n'\rangle F_{n'}(z-ld)F_n(z-ld) \tag{3.1.24}$$

where

$$\pi_{nn'}(\vec{q},\omega) = 2\sum_k \frac{f(E_{n;\vec{k}_t+\vec{q}}) - f(E_{n,\vec{k}_t})}{E_{n;\vec{k}_t+\vec{q}} - E_{n,\vec{k}_t} - \hbar\omega} \tag{3.1.25}$$

is the irreducible polarizability. Here, the relation

$$E_{n,\vec{k}_t} = E_n + (\hbar^2 k^2)/(2m_e^*) \tag{3.1.26}$$

has been taken, where E_n is the energy of the n-th quantized z-motion level. The induced potential V^{ind} is related to the density response δn by Poisson's equation

$$\nabla^2 V^{ind} = -(4\pi e^2)/(\varepsilon_s)\delta n \tag{3.1.27}$$

where ε_s is the background dielectric constant of the superlattice system. Solving this equation one obtains

$$V^{ind} = (2\pi e^2)/(\varepsilon_s q) \int_{-\infty}^{+\infty} dz' \exp[-q|z-z'|]\delta n(\vec{q},\omega,z') \tag{3.1.28}$$

The matrix elements of V^{ind} are

$$\langle nl|V^{ind}|n'l\rangle = \sum_{mm'l'} \pi_{mm'} V_{nn'}^{mm'} \langle m|V_{l'}|m'\rangle \tag{3.1.29}$$

where

$$V_{nn'}^{mm'} = (2\pi e^2)/(\varepsilon_s q) \int_{-\infty}^{+\infty} dz\, dz' F_n(z)F_{n'}(z)\exp[-q|z-z'+ld-l'd|] \cdot$$

$$\cdot F_m(z')F_{m'}(z') \tag{3.1.30}$$

Now taking the nkl,n'k'l' matrix elements of the potential $V=V^{ext}+V^{ind}$ and using the Eq. (3.1.29) yields

$$\langle n|V_l|n'\rangle = \langle n|V_l^{ext}|n'\rangle + \sum_{mm'l} \pi_{mm'}\langle m|V_{l'}|m'\rangle V_{nn'}^{mm'} \tag{3.1.31}$$

Now one can assume here for simplicity, the electric quantum limit. This means that only the lowest quantized z-motion level is occupied in each layer. In that case n'=0 and when defining $\chi_{m0} = \pi_{m0} + \pi_{0m}$ one can rewrite the last equation as

$$\langle n|V_l|0\rangle = \langle n|V_l^{ext}|0\rangle + \sum_{ml'} \chi_{m0} \langle m|V_{l'}|0\rangle V_{nm} \quad (3.1.32)$$

where $V_{nm} = V_{n0}^{m0}$. This equation gives the linear response of the superlattice to an external potential V^{ext} given by the Eq. (3.1.20).

For the collective modes to occur $V^{ext} = 0$ while $V \neq 0$. Thus, Eq. (3.1.32) becomes

$$\langle n|V_l|0\rangle = \sum_{ml'} \chi_{m0} \langle m|V_{l'}|0\rangle V_{nm} \quad (3.1.33)$$

In view of the translational symmetry of the superlattice system the last equation may be written in the form

$$\langle n|V_l|0\rangle = \exp(ik_z ld)\langle n|V_0|0\rangle \quad (3.1.34)$$

which defines the relative phase of electron density oscillations in adjacent planes. Assuming now that all relevant z,z' in the integral (3.1.30) are much smaller than ld,l' d, i.e. $(z,z') \ll |l-l'|d$, unless l=l', one may write the condition for collective modes as

$$\det|\delta_{nm} - \chi_{m0}[V_{nm}(q) + S_-\hat{V}_{nm}(q) + S_+\hat{V}_{nm}(q)]| = 0 \quad (3.1.35)$$

where

$$V_{nm}(q) = (2\pi e^2)/\varepsilon_s q \int dz\, dz'\, F_n(z)F_0(z)\exp[-q|z-z'|]F_m(z')F_0(z') \quad (3.1.36)$$

and $\hat{V}_{nm}(q)$ differs from $V_{nm}(q)$ only in having $\exp[-q|z-z'|]$ replaced by $\exp[-ik_z(z-z')]$. The functions $S_\pm(q,k_z)$ are defined by

$$1 + S_\pm(q,k_z) = [1-\exp(-qd)\exp(\pm ik_z d)]^{-1} \quad (3.1.37)$$

It is convenient to introduce $S(q,k_z) = 1 + S_+ + S_-$ where

$$S(q,k_z) = \frac{\sinh qd}{\cosh qd - \cos k_z d} \quad (3.1.38)$$

For purely two-dimensional layers $n = m = 0, F^2(z-ld) \to \delta(z-ld)$, $V_{nm} = \hat{V}_{nm} = (2\pi e^2)/(\varepsilon_s q)$, $\chi_{m0} = \pi_{00}$, and Eq. (3.1.35) becomes

$$1 - (2\pi e^2)/(\varepsilon_s q)\pi_{00}(q,\omega)\left[\frac{\sinh qd}{\cosh qd - \cos k_z d}\right] = 0 \quad (3.1.39)$$

which is precisely the dispersion relation obtained in |37|.

In general, it is difficult to obtain exact dispersion relations from Eq. (3.1.35). This problem may, however, be simplified by making two approximations |36|. First, the determinant may be truncated by neglecting all off-diagonal elements, essentially assuming no mixing between different intersubband excitations. Secondly, all quantities may be expanded to $O(q^2)$. In doing so, the quantity qd becomes fixed so that S can be treated as a constant. The following dispersion relation is then obtained

$$\omega^2 = \Omega_{h0}^2(1 + \alpha_{nn}) - \Omega_{h0}^2 \mu_{nm} Sq + [m^{-1}\hbar\Omega_{n0}(1 + \alpha_{nn}) + + V_F^2 (\alpha_{nn}^{-1} + 3/4) - \Omega_{n0}^2 \gamma_{nn}]q^2 \quad (3.1.40)$$

where

$$\alpha_{nn} = -(4 e^2\pi N_s)/(\hbar\Omega_{n0}\epsilon_s)\int dz\, dz' F_n(z)F_0(z)|z-z'|F_n(z')F_0(z')$$

$$\mu_{nn} = (4\pi e^2 N_s)/(\hbar\Omega_{n0}\epsilon_s)|z_{n0}|^2$$

$$\gamma_{nn} = (4\pi e^2 N_s)/(\hbar\Omega_{n0}\epsilon_s 6)\int dz\, dz' F_n(z)F_0(z)|z-z'|^3 F_n(z')F_0(z')$$

N_s is here the electron concentration per unit area in each layer, and $h\Omega_{n0} = E_n - E_0$ is the subband separation.

The dispersion relation (3.1.40) may be now examined in various limits. In the strong-coupling limit (qd<<1), where the layers of the superlattice are very close together, there are two cases to consider:

Case 1 is characterized by the relations (qd<<1) and $k_z \neq 0$. In this case the phase of the density oscillation in adjacent layers changes in a definite way, and consequently $S(q,k_z) \approx qd|1-\cos k_z d|^{-1}$, so that

$$\omega^2 \approx \Omega^2(1 + \alpha_{nn}) + \left[\frac{-\Omega_{n0}^2 \mu_{nn} d}{1 - \cos k_z d} + m^{-1}\hbar\Omega_{no}(1 + \alpha_{nn}) + + V_F^2(\alpha_{nn}^{-1} + 3/4) - \Omega_{n0}^2 \gamma_{nn}\right]q^2 \quad (3.1.41)$$

Case 2 is characterized by (qd<<1) and $k_z = 0$. In this case the electron density oscillations are in phase in each layer, and consequently $S(q,k_z) \approx 2/(qd)$, so that

$$\omega^2 \approx \Omega_{n0}^2(1 + \alpha_{nn} - 2d^{-1} \mu_{nn}) + O(q^2) \quad (3.1.42)$$

Here the terms of $O(q^2)$ are unchanged. In this case the layers are so close together that charge oscillations in adjacent layers are able to decrease the polarization charge, and thus decrease the depolarization shift.

In the weak-coupling limit (qd>>1) the structure function $S(q,k_z) = 1$ and Eq. (3.1.40) reduces to the dispersion relation for a single quantum well |45|. In this limit each layer would be expected to support its own intersubband collective mode.

3.2. Energy states of confined carriers in single quantum wells and in type I superlattices

The basic element of a superlattice is the single quantum well. It is created in compositional superlattices by the heterostructures of constituent semiconductors, and in doping superlattices by differently doped neighbouring layers of the host semiconductor. Sometimes the term superlattice is used to refer to a periodic structure which can be considered as a set of independent quantum wells |32|,|46|. In such a structure the barrier layers are thick enough to prevent any coupling between the wave functions describing the energy states of the carriers confined to the individual quantum wells. In fact, such periodic structures do not exhibit properties typical for superlattices, showing rather the properties of a single quantum well in multiplication. Therefore, such structures are called multi-quantum-well (MQW) structures. They can be treated as superlattices in the limit of infinitely thick barriers. As a matter of fact, the first direct experimental observation of bound-electron and bound-hole states of rectangular potential wells has been reported for MQW GaAs-$Al_xGa_{1-x}As$ structures consisting of fifty GaAs layers separated by $Al_xGa_{1-x}As$ layers thicker than 25 nm |46|.

It is worthwhile to discuss first the quantum states in single, separated wells before proceeding with the theory of superlattices. The theory is simpler and the separation in energy between the bound-carrier states defines the depth of the wells, and thus the amplitude of the superlattice potential, the basic parameter for the superlattice.

The simplest quantum well which appears in superlattices is the rectangular, or square-shaped potential well, of the type I superlattices |32|. If a particle is completely confined to a layer of thickness L_z by an infinite potential well, then the energies of the stationary bound states, resulting from the solution of the time-independent Schrödinger equation are equal to

$$E = E_{ni} + [\hbar^2/(2m)](k_x^2 + k_y^2) \qquad (3.2.1)$$

where

$$E_{ni} = [(\hbar^2\pi^2)/(2m_i)](n/L_z)^2 , \quad n = 1,2,3 \ldots \infty \qquad (3.2.2)$$

k_x, k_y are here the wave vectors of the particle in the directions parallel to the well walls. The motion of the particle in the z-direction is quantized, while in the x and y directions the particle moves classically. One can see from the last equation, that the particles confined to the quantum well exhibit the so-called quantum size effect, i.e., the dependence of the energies E_n on the width L_z of the well. This effect plays an important role in superlattices, because it defines the effective band gap of these structures.

In reality, the potential wells in type I superlattices have finite depths and the above presented solutions of the Schrödinger equation are inadequate for a quantitative analysis of the quantized energy states of the confined particles.

3.2.1. Real quantum wells

The solution of the time-independent Schrödinger equation for a one-dimensional finite square-well potential is a textbook quantum mechanical problem (see, e.g. |48|,|49|). In the case of the $GaAs-Al_xGa_{1-x}As$ MQW structure, this problem has been solved for electrons and for two sets of holes, the heavy and the light holes, which are associated with the ±3/2 and the ±1/2 Γ_8 valence bands, respectively. To this end, for each of these types of carriers the Schrödinger equation has been solved separately with the particle mass m_i being replaced by the relevant effective mass m^*, characteristic for GaAs, i.e., with $m_e^* = 0.0665\ m_0$ for electrons, $m_{hh}^* = 0.45\ m_0$ for heavy holes, and $m_{lh}^* = 0.08\ m_0$ for light holes |46|. It has been found that both the level spacing and the number of bound states decreases as the well depth is decreased, but the n = 1 state exists for all positive values of the depth. Thus, for all potential wells in the MQW structure, at least one bound state exists for each type of the carriers. This theoretically predicted behaviour has been confirmed experimentally |32|, what shows that the simple effective mass approximation is quite appropriate to describe the particles in finite square wells. More sophisticated microscopic calculations |4| have indeed confirmed that the accuracy of this approximation is satisfactory for the considered problem. The electronic structure of a single quantum well configuration of the GaAs-$Al_xGa_{1-x}As$ MQW structure (with only one valence band shown, for clarity) is illustrated in Fig. 3.3.

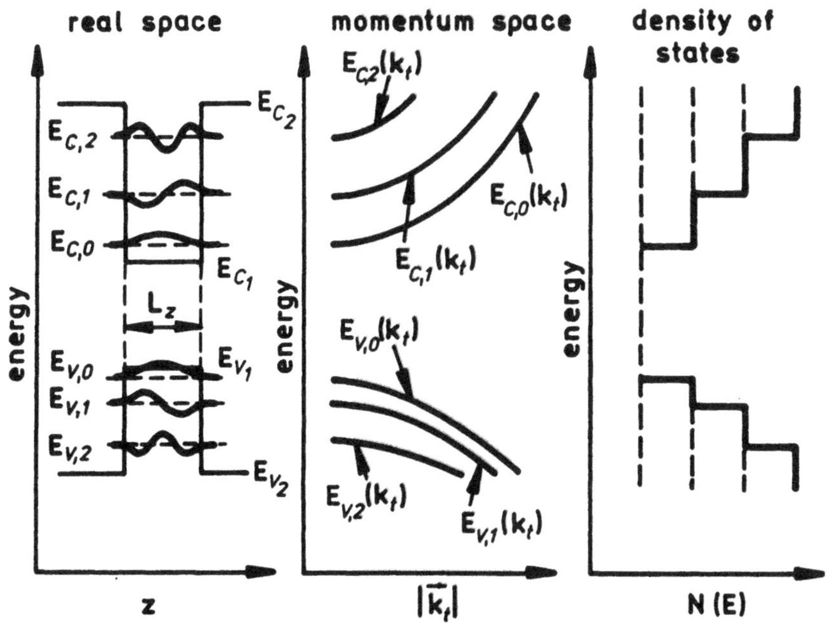

Fig.3.3. Schematic illustration (not to scale) of the electronic structure of a single quantum well configuration formed by a thin layer of GaAs sandwiched between two thick cladding layers of $Al_xGa_{1-x}As$. For clarity, only one valence band is assumed (after Ploog et al.,Ref. 3)

As it has already been mentioned before, the spatial distribution of doping impurities typical for the type I modulation doped superlattices (see,Fig. 1.7.) as well as for the doping superlattices (see, Fig. 1.6.) leads to the problem of triangular and parabolic quantum wells, respectively. In triangular wells the energy levels, as well as the density-of-states functions of the confined particles are varying approximately as the 2/3 power of the quantum number |50|,|51|, while in the parabolic wells both these quantities are uniformly spaced |52|, resembling the harmonic-oscillator spectrum of the energy levels.

A single parabolic quantum well can be realized, for example, by creating n-p^+-n or p-n^+-p doping structures |53|. Such a structure with highly doped n^+-layer of thickness z_D and doping concentration N_D embedded in a weakly doped p-type semiconductor with acceptor concentration N_A is shown schematically in Fig. 3.4a. The corresponding band profile of this structure at thermodynamical equilibrium (without bias) is shown in Fig. 3.4.b, with the impli-

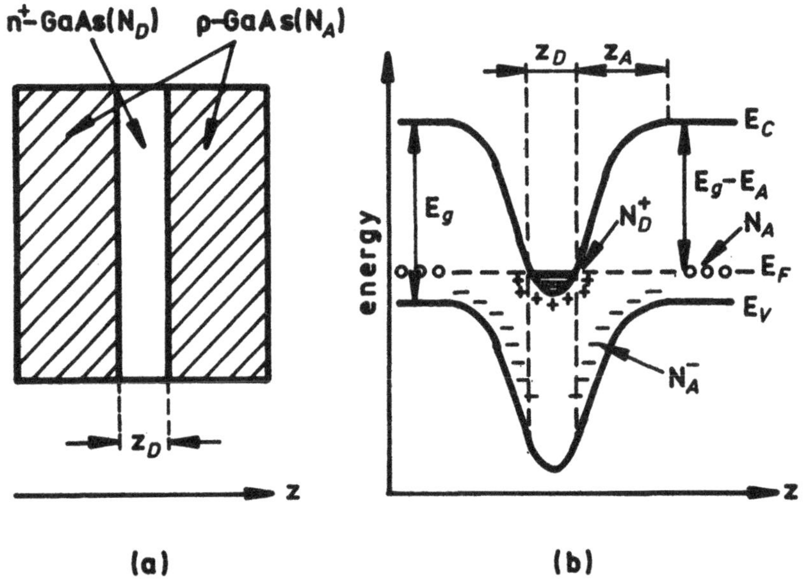

Fig. 3.4. Ultrathin n^+-GaAs layer doped with donors of concentration N_D embedded in a p-GaAs layer weakly doped with acceptors of concentration N_A shown schematically (a), and the space charge potential profile of the p-n^+-p-GaAs system in the thermodynamical equilibrium, shown in the real space (b) (after Döhler,Ref. 53).

cit assumption made, that the number of donors per unit area in the n^+-layer exceeds that of acceptors in the two negative space charge layers, i.e. $N_D z_D >> 2N_A z_A$. The length z_A of the space charge region of negatively charged acceptors below the Fermi level follows from

$$E_g - E_A \approx (2\pi e^2 N_A/\varepsilon) z_A^2 \qquad (3.2.3)$$

[compare with Eq. (1.3.1)]. The neutrality condition then yields a density of conduction electrons per unit area in the n^+-layer of

$$n(z) = N_D z_D - 2N_A z_A \qquad (3.2.4)$$

These electrons are subject to the steep and strong space charge potential of the p-type space charge layers for motion in z-direction. Therefore, the energy spectrum is quantized with respect to the z-component. The motion parallel to the n^+-layer axis, however, is only influenced by the relatively weak potential fluctuations which originate from the random distribution of the impurities, in particular of the donors in this layer. The long range part of these fluctuations becomes increasingly well screened with increasing electron density n(z), therefore, at sufficiently large values of n(z) electrons near the Fermi level can be considered as a 2DEG layer. The effective mass m_e^* and the momentum parallel to the layer axis $\vec{k}_t$ of these electrons define the energies of the stationary quantum states in the parabolic potential well, which are described by Eq. (3.2.1) with

$$E_{ni} = (n - 1/2)(\hbar/z_D)(2V_0/m_e^*)^{1/2}, n = 1,2,3...\infty \qquad (3.2.5)$$

V_0 is here given by Eq. (1.3.1) in which d is replaced by $2z_D$ |3|.

Quantum wells with parabolic potential profiles can also be generated in a GaAs-$Al_xGa_{1-x}As$ heterojunction structure. This can be done by alternate deposition of thin undoped layers of GaAs and $Al_xGa_{1-x}As$ of varying thicknesses. The relative thicknesses of the $Al_xGa_{1-x}As$ layers have to increase quadratically with distance from the well centres, while that of the GaAs layers must decrease. The average layer thicknesses of approximately 1 nm are usually employed in order to permit the GaAs layers to be sufficiently thick to produce surface smoothing and cleaning, while still allowing ample electron and hole tunneling to average the effective potentials to parabolic profiles.

In such a structure realized in practice by using the MBE growth technique |52| each well contained 20 layers of $Al_xGa_{1-x}As$ and 21 layers of GaAs. The thickness of the N-th layer of $Al_xGa_{1-x}As$ from the centre of the well was $[(N-0.5)/10]^2(L_z/20)$. The $Al_xGa_{1-x}As$ layers were centred at distances $(N-0.5)(L_z/20)$ from the well centre, and the remaining material was GaAs.

The energies of the stationary bound states for infinitely deep parabolic wells are given in this case by Eq. (3.2.1.) with

$$E_{ni} = (n - 1/2)\hbar\omega_{0i} = (n - 1/2)\hbar[K_i/m_i^*]^{1/2} \quad , \ n=1,2,3\ldots\infty \qquad (3.2.6)$$

K_i is here the curvature of the well for the particle with the effective mass m_i^* |52|. It may be determined by the potential height of the finite parabolic well at $z = \pm(L_z/2)$. In the case of the material system GaAs-$Al_xGa_{1-x}As$, the potential height is equal to the band-edge discontinuity $Q_i\Delta E_g$ for the i-th particle well, with ΔE_g being the total energy gap discontinuity between the GaAs at the bottom of the wells and the $Al_xGa_{1-x}As$ at the top of the wells. $Q_e\Delta E_g$ then stands for the conduction band-edge discontinuity ΔE_C and $Q_h\Delta E_g$ for the valence band-edge discontinuity ΔE_V. Equation (3.2.6) may thus be written in the form |52|

$$E_{ni} = 2(n - 1/2)(\hbar/L_z)[2Q_i\Delta E_g/m_i^*]^{1/2} \qquad (3.2.7)$$

This equation is, of course, only an approximation since the finite well heights should be taken into account as well as the dependence of the effective mass on the $Al_xGa_{1-x}As$ alloy composition x. Nevertheless, it gives reasonable agreement (in the range of a few percent error) with the photoluminescence experimental data, provided the characteristic material parameters for the GaAs quantum wells are equal to: $Q_e = 0.57$, $Q_h = 0.43$, $m_e^* = 0.0665\ m_o$, $m_{hh}^* = 0.34\ m_o$ and $m_{lh}^* = 0.094\ m_o$ |52|,|54|.

The physical quantities used for describing the stationary bound states in quantum wells, such as effective masses, energy gaps, and others, have their usual meanings primarily over a spatial region that spans many atoms. At real interfaces of the wells, specially at heterointerfaces between two different materials, at least one layer of atoms exists whose bonding environment differs from that in either of the adjoining layers. The main physical properties do not make, therefore, an abrupt transition from their value in one material to their value in the other, but can be expected to have a region in which the values are changing. The effect of a graded interface on a quantum well has been examined in |47|,|55|, by using both, a continuum model with rapidly varying parameters of the effective mass Hamiltonian of the BenDaniel-Duke form |56|, and a microscopic model that uses a tight-binding formalism.

It has been found that the two models give similar results evidencing that the graded interfaces, few atomic layers thick, have only a minor effect on energy levels in the GaAs-$Al_xGa_{1-x}As$ square wells.

While slightly graded interfaces do not influence seriously the quantum states of the confined charge carriers, an external constant electric field of the strength in the range of 10-50 kV/cm applied in the direction perpendicular to the walls of the quantum well, perturbes significantly the free electron and hole bound states in the well |57|. This effect is clearly seen in optical experiments performed on MQW structures. The photoluminescence associated with free exciton recombination |58| can be almost quenched by the longitudinal electric field $\xi_\perp$. Also a quadratic in $\xi_\perp$ shift of the luminescence peak positions may be observed with increasing $\xi_\perp$ |59|. This result has been interpreted in terms of electric-field-induced dissociation of the excitonic electron-hole pairs. In electro-absorption experiments performed on MQW structures |60| the excitonic absorption peaks are red-shifted when a longitudinal electric field is applied on the MQW's via a p-i-n diode. This red shift is interpreted in terms of a variation of the free electron and free hole confinement energies in the well, occurring in a close dependence on $\xi_\perp$ (this effect is also known under the term "Stark shift"). Together with the Stark shift a modification of the exciton binding energy occurs. The excitonic structures become increasingly broadened with increasing $\xi_\perp$. Electro-reflectance measurements |61| display similar features, i.e., Stark red shift of electroreflectance structures, and broadening of these structures at high fields. The theoretical analysis of the effect which a constant electric field exerts on the energy position of the exciton ground state in semiconductor single quantum wells of, e.g., GaAs-$Al_xGa_{1-x}As$ MQW structures, where electron and hole wave functions are mostly localized within the same layer at zero electric field, requires variational calculations to be applied to this problem |62|,|63|. It has been found that the exciton binding energy is little affected by $\xi_\perp$ in narrow QW's whereas in structures of $L_z \geqslant 10$ nm a significant reduction of this energy takes place. At low fields a quadratic Stark shift has been found, which is followed by a slower $\xi_\perp$ dependence characteristic of carrier accumulation on each side of the QW. Finally, at very large $\xi_\perp$ the carrier sweep out of the QW takes place which should manifest itself by a vanishing of the exciton binding energy. These theoretical results are in a reasonable agreement with the described experimental observations |59|.

So far, we have discussed exclusively the energies of the stationary states of free carriers confined to quantum wells. However, luminescence experiments performed on slightly doped GaAs quantum wells have exhibited structures with intensities strongly dependent on temperature, which are energetically located below the spectral lines of free-carrier recombination related to electronic transitions between the energy states of the confined carriers. This has evidenced that also localized impurity states can occur in the wells |64|. The first step towards understanding the nature of these impurity levels has been made by Bastard who has used a variational procedure in order to analyse the localized states of a hydrogenic impurity in a quantum well |65|.

Assuming that there are no image forces acting inside the single, separated quantum well, and that the carrier kinematics in the well is described by a one-band spherically symmetric effective-mass Hamiltonian, one obtains the problem to be solved as described by the following one-impurity Hamiltonian |65|

$$H = p^2/(2m^*) - e^2\varepsilon^{-1}\left[\rho^2 + (z-z_i)^2\right]^{-1/2} + V(z) \qquad (3.2.8)$$

where $V(z)$ is the potential-energy barrier which confines the carriers in the well (the depth of the well), $\rho^2 = x^2+y^2$ is the distance, measured from the impurity site in the plane parallel to the well walls, and z_i is the co-ordinate of the impurity site along the axis perpendicular to the well walls. The origin of the co-ordinate system is located at the centre of the well. Considering for the sake of definitness only the donor levels [it is clear that Eq. (3.2.8) also applies to acceptor states provided H is understood as the hole Hamiltonian], the eigenstates of (3.2.8) without the impurity potential, are given by

$$\langle\vec{r}|n\vec{k}_t\rangle = [2/(L_zS)]^{1/2}\exp(i\vec{k}_t\vec{\rho}) \begin{cases} \cos k_z z \ , & n \text{ odd} \\ \sin k_z z \ , & n \text{ even} \end{cases} \qquad (3.2.9)$$

where L_z and S are the well thickness and the well wall surface, respectively, $\vec{k}_t = (k_x,k_y)$, and $k_z = (n\pi)/L_z$ with $n\geq1$.

The eigenenergies corresponding to the wave function (3.2.9) are given by Eqs. (3.2.1) and (3.2.2) with m replaced by the carrier effective mass m^*. Because the transverse and longitudinal variables (ρ,z) do not separate, Eq. (3.2.8) cannot be solved exactly. In the variational procedure applied in |65| the trial wave function used has the form

$$\langle\vec{r}|\psi\rangle = N \cos k_1 z \exp\{-\lambda^{-1}[\rho^2 + (z-z_i)^2]^{1/2}\} \qquad (3.2.10)$$

if $|z|\leq L_z/2$. This function is equal to 0 outside the well (if $|z|>L_z/2$). λ is here the variational parameter and N is a normalization coefficient. Denoting by $E(L_z,z_i)$ the eigenenergies of (3.2.8) the impurity binding energy $E_i(L_z,z_i)$ will be equal to

$$E_i(L_z,z_i) = (\hbar^2\pi^2)/(2m^*L_z^2) - E(L_z,z_i) \qquad (3.2.11)$$

The trial wave function chosen is an exact solution of the Hamiltonian (3.2.8) in both limits, $L_z = 0$ and $L_z = \infty$. For $L_z = \infty$ and $z_i = 0$, Eq. (3.2.10) reduces to the ground state of the three-dimensional hydrogen atom

$$\langle\vec{r}|\psi\rangle = (\pi\lambda^3)^{-1/2}\exp[-\lambda^{-1}(\rho^2 + z^2)^{1/2}] \qquad (3.2.12)$$

with $\lambda = a_0^* = (\varepsilon\hbar^2)/(m^*e^2)$ and $E_i(\infty,0) = R_0^* = (m^*e^4)/(2\varepsilon^2\hbar^2)$, whereas if $z_i = \pm L_z/2$ and $L_z = \infty$ the $2p_z$ hydrogenated state is obtained $[\lim_{L_z\to\infty} E_i(L_z,\pm L_z/2) = R_0^*/4]$. For $L_z = 0$ the expression (3.2.10) reduces to the ground state of the two-dimensional hydrogen atom

$$\langle\vec{\rho}|\psi\rangle = [2/(\pi\lambda^2)]^{1/2}\exp(-\rho/\lambda) \qquad (3.2.13)$$

with $\lambda = a_0^*/2$, corresponding to a binding energy $[\lim_{L_z\to0} E_i(L_z,z_i) = 4R_0^*]$. It should be noticed that the trial function (3.2.10) is one of the simplest which is compatible with the perfect confinement, $\langle\vec{r}|\psi\rangle = 0$ if $|z|\geq L_z/2$. With this function the eigenenergies of the Hamiltonian (3.2.8) can be obtained in a closed form

$$E(L_z,z_i) = (\hbar^2k_1^2)/(2m^*) + \hbar^2/(2m^*\lambda^2) - (e^2\pi N^2\lambda^2)/(2\varepsilon)\Big[1 + \frac{\cos(2k_1z_i)}{1+k_1^2\lambda^2} - \frac{k_1^2\lambda^2}{1+k_1^2\lambda^2}\exp(-L_z/\lambda)\cosh(2z_i/\lambda)\Big] \qquad (3.2.14)$$

Extremalizing this expression with respect to λ one obtains the impurity binding energies $E_i(L_z,z_i)$ as a function of the well thickness for a given z_i, or as a function of the impurity position z_i along the axis perpendicular to the well walls, for a given thickness L_z. It has been found |65| that E_i are decreasing functions of the well thickness, reaching R_0^* and $R_0^*/4$ in the infinite thickness limit, for $z_i = 0$ and $\pm L_z/2$, respectively. At constant well thickness L_z, the $E_i(L_z,z_i) = E_i(L_z,-z_i)$ is maximum if the impurity is located at the centre of the well ($z_i=0$) and reaches its minimal value if $z_i = \pm L_z/2$. These findings have been confirmed experimentally by photoluminescence investigations for Si donors in GaAs quantum wells |66|,|67|.

The problem of hydrogenic impurity levels has been studied further by extending the analysis to higher-order two-dimensional subbands |68|,|69|. It has been shown, that in the quantum wells there exist also resonant impurity states, i.e., states which are falling in the continuum of the energy states of lower subbands.

The variational method has also been applied to calculate the acceptor binding energies in GaAs-$Al_xGa_{1-x}As$ quantum wells |70|,|71|. Because the quantum well potential reduces the symmetry of the GaAs system from T_d to D_{2d}, the bulk Γ_8 acceptor ground state splits into Γ_6 and Γ_7 states. The Γ_6 state is predominantly derived from the heavy-hole subband, while the Γ_7 state is related to light-hole subband. Accordingly, in these calculations the coupling of the top four valence bands of both materials creating the wells have been included, and a multiband effective-mass approximation has been used. The results were qualitatively identical with the results of Ref. 65. These results indicated, in a very good agreement with the experimental data, that again the maximum binding energy occurs when the acceptor is located at the centre of the well, and the minimum binding energy occurs when the acceptor is at the edge of the well. For wider wells the acceptor binding energy is much more position dependent then for narrower wells. Except for well widths smaller than ≈ 4 nm, the Γ_7 binding energy is greater than the Γ_6 binding energy.

Concluding the considerations related to impurity states in quantum wells, it should be emphasized, that in both cases, of donors and of acceptors, the binding energies exhibited by the impurities located at the centre of the wells are significantly larger than those of the same impurities when located in the bulk of the material used to create the well.

3.2.2. Lattice matched type I superlattices

In the type I superlattices, one is usually concerned with the square-well potential for both kinds of carriers, i.e., for electrons and for holes. Therefore, using a simply periodic square--well model for the potential distribution in the direction of the superlattice axis, the parameters of the superlattice subbands can be derived from solutions of the Schrödinger equation, based on the Kronig-Penney model of the one-dimensional crystal lattice |72|,|73|. To this end the effective mass of the electrons in the superlattice may be taken, in the first approximation, to be a suitable average of their effective masses in the two constituent materials |74|,|75|. This very crude approximation giving satisfactory qualitative information about the superlattice energy subbands, is unlikely to give quantitatively good results |76|. Consequently, more exact theories should be developed. All of these theories |17|,|77-79|, have been based on the effective mass approximation. We will follow here the procedure proposed in Ref.78, because it is simple enough, yet gives results which agree well with more complicated thight-binding calculations which do not employ the effective mass approximation |11|. In this procedure, a transfer matrix is used which gives a linear relation between envelopes of the wave functions and their first derivatives at the interface. This matrix is applied to the calculation of the subband structure of the type I, as well as of the type II, superlattices. However, we will discuss here the procedure only in relation to the GaAs-$Al_xGa_{1-x}As$ superlattice.

The band structures of the constituent compounds are well reproduced by a linear combination of s, p_x, p_y and p_z orbitals in the nearest neighbour (empirical) tight-binding approximation |80|. Especially the conduction band and the light-hole valence band can be described by a s-orbital on the anion site and a p_z-orbital on the cation site.

Let us now consider, in accordance with |78|, a junction of a semiconductor denoted by A occupying the left (z<0) half-space and a semiconductor B occupying the right (z>0) halfe-space. We will assume for simplicity, that they possess single band maximum or minimum which lies close in energy, and discuss energy levels close to such band extrema. The wave functions of A and B should match smoothly with each other at the interface (z=0). This does not necessarily mean that envelope functions F(z) are smooth across the interface. Nevertheless, if the wave function of one side far from the interface is given, that of the other side is uniquely determined for any kind of junctions. This fact can be described by the following relation between envelopes and their derivatives at z=0.

$$\begin{vmatrix} F_B(0) \\ a_0 \frac{\partial}{\partial z} F_B(0) \end{vmatrix} = T_{BA} \begin{vmatrix} F_A(0) \\ a_0 \frac{\partial}{\partial z} F_A(0) \end{vmatrix} \qquad (3.2.15)$$

where $T_{BA} = (t_{ij})$ is a 2x2 matrix, which is called-transfer matrix, and which contains all the information of the microscopic structure and properties of the interface. Since this transfer matrix is determined by quantities of atomic scale, it can be used together with effective-mass Schrödinger equations for electrons and for holes, to calculate the subband structure of a superlattice in the presence of weak slowly varying potentials arising, for example, from space charges.

There are various methods to construct T_{BA}. It should be noticed, however, that the four components of this matrix are not independent of each other. The electron flux conservation leads to the conclusion that T_{BA} can be chosen as real except for the common phase factor, and that its determinant should be equal to m_B^*/m_A^*, where m_B^* and m_A^* are the effective masses of electrons in semiconductors B and A, respectively. T_{BA} is also independent of k_x and k_y within the effective-mass approximation, which describes wave functions correctly only to the order of ka_0.

The simplest way to construct the transfer matrix is to calculate the transmission and reflection coefficients together with phase shifts of waves passing through the interface. Let us consider a one-dimensional tight-binding model of the interface created by a chain of atoms having single s-orbitals (s-scoupling bands). In this case

$$E_a c(n) - t\, c(n+1) - t\, c(n-1) = E\, c(n) \tag{3.2.16}$$

where E_a (E_a^A and E_a^B for A and B, respectively) is the atomic energy, t (= t_A, t_B or t_{AB}) is the transfer energy, and c(n) is the amplitude of the wave function at the n-th site. At the interface located between the N-th and the (N+1)-th sites

$$t_A c^A(N+1) = t_{AB} c^B(N+1) \tag{3.2.17}$$

and

$$t_{AB} c^A(N) = t_B c^B(N) \tag{3.2.18}$$

where $c^A(N+1)$ and $c^B(N)$ are the extrapolated values past the interface from left and right, respectively. Putting $c^A(N) = F_A(-a_0/2) = F_A(0) - [a_0 \frac{\partial}{\partial z} F_A(0)]/2$ one gets

$$T_{BA} = \begin{vmatrix} (1/2)(t_A/t_{AB} + t_{AB}/t_B) & (1/4)(t_A/t_{AB} - t_{AB}/t_B) \\ t_A/t_{AB} - t_{AB}/t_B & (1/2)(t_A/t_{AB} + t_{AB}/t_B) \end{vmatrix} \tag{3.2.19}$$

where the origin z=0 has been chosen at the centre of the line connecting the N-th and (N+1)-th atomic sites. The last equation shows that the connection between the envelopes becomes weak wher t_{AB} deviates from the geometric mean of t_A and t_B.

Similar considerations lead to a transfer matrix

$$T_{BA} = \begin{vmatrix} t_A/t_B & 0 \\ 0 & (t_{AB} t_A E_g^B)/(t_B^2 E_g^A) \end{vmatrix} \tag{3.2.20}$$

in the case of the interface created by two kinds of atoms, one having a s-like orbital and the other having a p_z-like orbital, when only the transfer between adjacent atoms is taken into account, and when the conduction bands of A and B are close in energy. The envelopes are now almost continuous across the interface (except in the case when the band parameters are extremely different from each other).

In the case of GaAs-$Al_xGa_{1-x}As$ one can get the transfer matrix by replacing t in (3.2.20) by $2V_{sp}$, where V_{sp} is the transfer energy The values of V^A_{sp} and $V^B_{sp} = V^{AB}_{sp}$ are determined so as to reproduce the bulk effective masses of GaAs and $Al_xGa_{1-x}As$, respectively. For x=0.1, the following values of the transfer matrix elements should be used |78|: $t_{11} = 1.021$, $t_{22} = 1.101$, and $t_{12} = t_{21} = 0$. The matrix is very close to the unit matrix, which means that the boundary condition stating that the envelopes are smooth across the interface is practically sufficient.

Using the boundary condition for the envelopes resulting from the Eq. (3.2.15), the band structure of the superlattice can be obtained by solving the Schrödinger equation in the effective-mass approximation for the superlattice potential having the one dimensional periodicity. This equation can be written for the conduction band electrons in the form

$$[-\hbar^2/(2m_e^*)\nabla^2 + V_{SL}(z)]\psi(r) = E\psi(r) \qquad (3.2.21)$$

To obtain the solution of this equation, we will follow the procedure given in |77|. Accordingly, we will make the following assumptions: (i) the interfaces between the layers are sharply defined, so any surface effects can be neglected, and the potential distribution may be considered to be an array of square wells, (ii) the electron mean free path is much larger than the superlattice period, so that the collision effects may be neglected, (iii) the effect of the atomic potentials is only to modify the effective-mass value of the electron corresponding to the E-k relation of the particular bulk material.

Let us consider an electron with energy E_A and wave vector $\vec{k}_A$ having the components k_x, k_y and k_1 in the region A (inside the well). Let the same electron, when it enters region B (the potential barrier) have energy E_B and wave vector $\vec{k}_B$ with the components k'_x, k'_y and k'_z (see, Fig. 3.5.a). Since energy is conserved in the transition and the wave functions at the boundary between two adjacent layers should be continuous, also the x and y components of the wave vector should be conserved, i.e., $k'_x = k_x$ and $k'_y = k_y$. Using the energy conservation condition and a parabolic E-k relation for

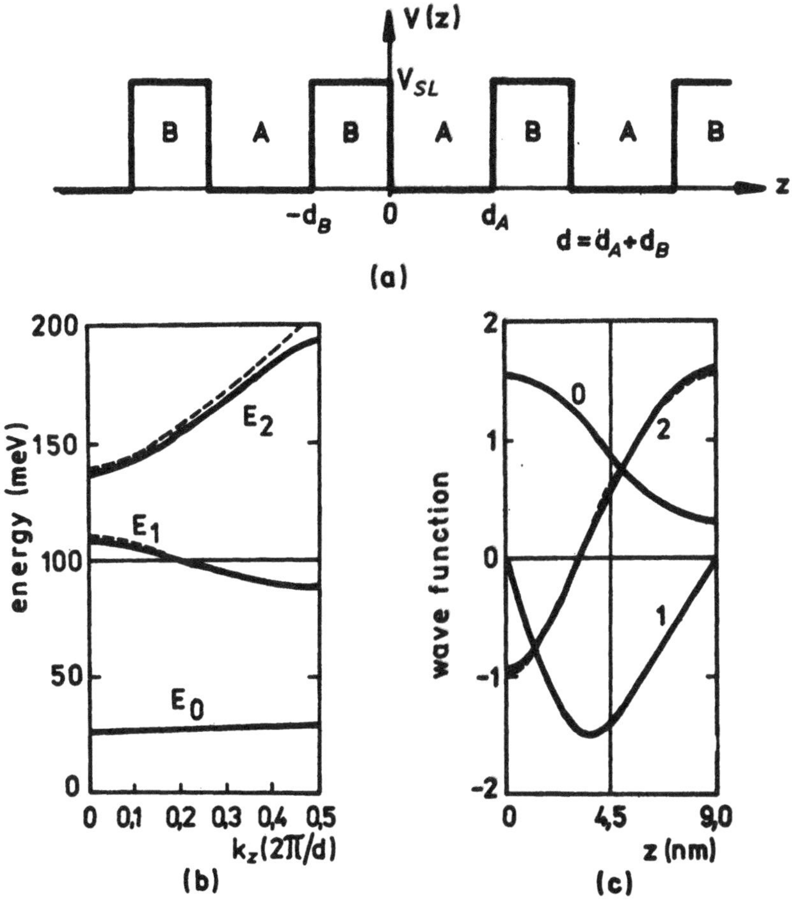

Fig.3.5. The Kronig-Penney one-dimensional periodic potential illustrating the band-edge variation in the conduction band of a GaAs-AlGaAs superlattice (a), the calculated subband energy versus k_z relation of the GaAs-$Al_{0.1}Ga_{0.9}As$ superlattice for $k_x = k_y = 0$, $d_A = d_B = 9$ nm and $V_{SL} = 0.1$ e V (b), and the envelope functions F(z), corresponding to the subband structure shown (c). The dashed lines are obtained by assuming that both the envelope and its derivative are continuous at the interface. The solid lines represent the results obtained with the boundary condition (3.2.15) [drawings (b) and (c) after Ando et al., Ref. 78].

the two materials, we obtain

$$[\hbar^2/(2m_A^*)](k_t^2 + k_1^2) = [\hbar^2/(2m_B^*)](k_t^2 + k_2^2) + V_{SL} \qquad (3.2.22)$$

where $k_t^2 = k_x^2 + k_y^2$. Since

$$k_1 = [(2m_A^* E)/\hbar^2]^{1/2} \qquad (3.2.23)$$

we obtain from Eq. (3.2.22)

$$k_2 = \{(2m_B^*/\hbar^2)[E-V_{SL}+E_t(1-m_A^*/m_B^*)]\}^{1/2} \qquad (3.2.24)$$

where E is the energy corresponding to the z-component of the wave vector in the potential well region, and $E_t = (\hbar^2 k_t^2)/(2m_A^*)$ is the transverse energy in the GaAs layer. We find from the above relation that k_2 is imaginary if $E<[V_{SL}-E_t(1-m_A^*/m_B^*)]$ and that its magnitude depends on E as well as on E_t, since $m_A^* \neq m_B^*$. Assuming now that $E_t(1-m_A^*/m_B^*) \ll V_{SL}$, the wave functions for the two regions may be written as

$$\psi_A(z) = A\exp[i(k_1-k)z] + B\exp[-i(k_1+k)z] \qquad (3.2.25)$$

$$\psi_B(z) = C\exp[(k_{2i}-ik)z] + D\exp[-(k_{2i}+ik)z] \qquad (3.2.26)$$

where

$$k_{2i} = [(2m_B^*/\hbar^2)(V_{SL}-E)]^{1/2}, \quad \text{for } 0<E<V_{SL} \qquad (3.2.27)$$

The constants A,B,C,D are to be chosen so that ψ and $\frac{d\psi}{dz}$ are continuous at z=0 and $z=d_A$ (at the boundaries of the well), and by the periodicity required of $\psi(z)$, according to which the value of this function at $z=d_A$ must equal that at $z=-d_B$. Thus we have the four linear homogeneous equations:

$$A + B = C + D \qquad (3.2.28)$$

$$i(k_1-k)A - i(k_1+k)B = (k_{2i}-ik)C - (k_{2i}+ik)D \qquad (3.2.29)$$

$$A\exp[i(k_1-k)d_A] + B\exp[-i(k_1+k)d_A] = C[\exp -(k_{2i}-ik)d_B] + D\exp[(k_{2i}+ik)d_B] \qquad (3.2.30)$$

$$i(k_1-k)A\exp[i(k_1-k)d_A] - i(k_1+k)B[\exp -(k_1+k)d_A] = (k_{2i}-ik)C\exp[-(k_{2i}-ik)d_B]-(k_{2i}+ik)D\exp[(k_{2i}+ik)d_B] \qquad (3.2.31)$$

This set of equations has a solution only if the determinant of the coefficients vanishes. This means that the solution exists only for energies which fulfil the following dispersion relation

|73|.

$$[(k_{2i}^2-k_1^2)/(2k_1k_{2i})]\sinh(k_{2i}d_B)\sin(k_1d_A)+\cosh(k_{2i}d_B)\cos(k_1d_A) = \cos(kd) \quad (3.2.32)$$

As the cosine term on the right side can have values only between +1 and -1, only those values of k_1d_A and $k_{2i}d_B$ are allowed, for which the left side falls in this range. The allowed ranges of k_1d_A and $k_{2i}d_B$ correspond through the relations (3.2.23) and (3.2.27) to the allowed ranges of the energy E of the electrons.

The calculated subband energy versus k_z relations and the envelope functions corresponding to the subband structure for a GaAs-$Al_{0.1}Ga_{0.9}As$ superlattice for $k_x=k_y=0$, $d_A=d_B$ = 9 nm and V_{SL} = 0.1 eV are shown in Fig. 3.5.b and c. For comparison, the results obtained with the boundary condition expressed by Eq. (3.2.15)(solid lines) and when assuming that both the envelope and its derivative are continuous at the interface (dashed lines) are shown on the same drawings.

The effective mass approximation has also been used for subband structure calculation in the modulation doped n-type GaAs-AlGaAs superlattice |81|. The motion of electrons parallel to the layers has been described by a free electron state with the effective mass $m^* = 0.068\ m_o$ in a medium with static dielectric constant $\varepsilon = 12.9$. For the motion perpendicular to the layers, a sequent square well potential has been used, where the $Al_xGa_{1-x}As$ layers created the barriers with a height V_{SL} = 0.3 eV. All the donors have been assumed to be ionized and replaced by a uniform positive charge distribution. Further the potential arising from the electrons themselves has been included self-consistently in the Hartree approximation.

In order to solve the relevant Schrödinger equation in the superlattice axis direction, the periodic part of the electron wavefunction and the potential have been expanded in terms of Fourier series. The problem has been reduced to solving the infinite simultaneous equations of the Fourier components, for which in the actual calculations the infinite series have been truncated at certain components in order to obtain convergent results. Fig. 3.6a shows a calculated result of the electron density distribution, potential energies, and energy spectra in the modulation

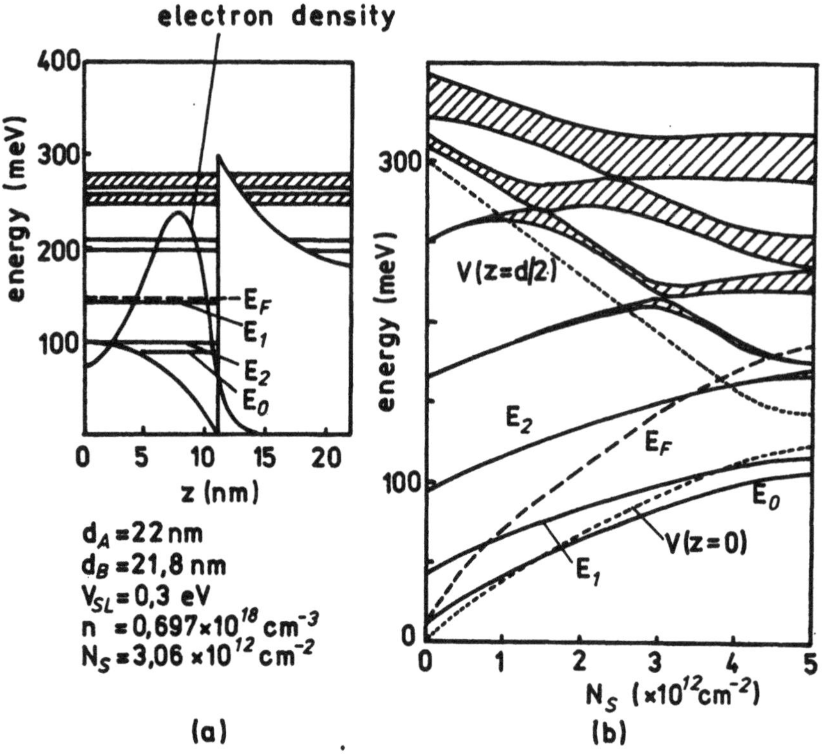

Fig.3.6. Calculated energy levels, density distribution of electrons, and the self-consistent potential for a modulation doping GaAs-n$^+$-$Al_xGa_{1-x}As$ superlattice. The subband widths are described by hatches (a). The same energy spectra as a function of the electron concentration, with the Fermi energy and the potential energy at z=0 and z=d/2 also shown (b) (after Mori et al., Ref. 81).

doping case (compare with Fig. 1.7). The z axis has been chosen in the superlattice direction, while the xy plane is located at the centre of the GaAs layer. N_S and n are the electron concentrations in a unit area and in a unit volume, respectively, with the relation $N_S = nd$, where $d = d_A + d_B$. Couplings between adjacent layers are weak for electrons in the three lowest subbands. The increase and decrease of the potential energy at z=0 and z=d/2, respectively are very large, and electrons are pushed toward the interface strongly if the band bending effect is included. Consequently, the lowest and the next lowest subbands are already close to two accumulation layer wave functions, which are coupled into bonding and antibonding levels. Fig. 3.6b shows the calculated energy spectra of the considered superlattice as a function of the electron concentration. Because of the band bending the energy of levels bounded in the GaAs layer increases with N_S, and the bottom of the ground subband is especially influenced and approaches to the bottom of the first excited subband. Around the electron concentration, where they cross the line corresponding to the potential at z=d/2, widths of the subband increase and their characters turn into three-dimensional. Above $N_S \approx 4 \cdot 10^{12}$ cm^{-2} the electron density distribution becomes extended as it begins to have a considerable amount in the $Al_xGa_{1-x}As$ layer.

3.2.3. Strained-layer superlattices

The electronic properties of strained-layer superlattices (SLS) are in general different from those of lattice matched superlattices due to the appreciable layer strains. The individual layers in the SLS match up lattice constants in the planes parallel to the interfaces by compression or expansion in these planes. Due to the Poisson effect, these layers are also expanded or compressed in the direction perpendicular to the interfacial planes. The resulting strains in the SLS layers consist of both hydrostatic and uniaxial components. The lattice constant of the SLS in the planes parallel to the interfaces is given by |82|

$$a_{0t} = \begin{cases} a_{01}[1 + (fG_2d_2)/(G_1d_1+G_2d_2)] \\ \text{or} \\ a_{02}[1 - (fG_1d_1)/(G_1d_1+G_2d_2)] \end{cases} \tag{3.2.33}$$

where a_{01} and a_{02} are the unstrained lattice constants of the superlattice constituent materials, d_1 and d_2 are the individual layer thicknesses, G_1 and G_2 are the shear moduli of the two materials, and $f = \Delta a_0/a_0$ is the lattice mismatch of the unstrained layers. From the last equation in can be deduced, that the ratios of the components of strain parallel to the interfaces for the two layers are inversely proportional to the ratio of the layer thicknesses, so that the thinner of the two types of layers has the greater layer strain. The limit thicknesses of the layers are determined, from the one side, by the one atomic monolayer thickness, and from the other, by the so-called critical thickness d_{crit}. The latter is the thickness at which it becomes energetically favourable for misfit dislocations to be generated |83|.

The first theoretical study of the electronic properties of a SLS has been performed for the GaAs-$GaAs_{0.2}P_{0.8}$ superlattice |84|. A semi-empirical tight-binding model with an extended unit cell |8|, has been employed. In this model the tight-binding parameter set for the strained materials includes both first and second nearest neighbour interactions to provide an adequate description of the lowest bulk conduction bands and includes spin-orbit parameters |85| to provide an adequate description of the strain splitting of the valence bands. Values for the tight-binding parameters of unstrained GaAs and GaP were obtained by fitting the tight-binding band structures to pseudopotential results |86|. The parameters for unstrained $GaAs_{0.2}P_{0.8}$ were obtained from weighted averages of the GaAs and GaP parameters. Strain was included in the model by adjusting the first nearest neighbour parameters for GaAs and GaP to produce the experimentally observed shifts and splittings of band energies under hydrostatic and uniaxial strain. Again the corresponding parameter shifts for $GaAs_{0.2}P_{0.8}$ were obtained from weighted averages of the shifts for GaAs and GaP. The GaAs/GaP band offsets were taken to be such that the unstrained GaAs Γ conduction band minimum line up with the unstrained GaP X conduction band minima. Offset values for the unstrained GaAs/ /$GaAs_{1-x}P_x$ system were obtained by assuming that the energy difference of the Γ conduction band minima for these materials varies linearly with the composition x. The strain dependence of the band offsets was taken to be due to the theoretical strain dependence of the ionization potential |87|. Although the numerical values for the superlattice energy gap in general depend on the values used for the band offsets, the qualitative features of the results obtained in |84| were not sensitive to the exact offset values used.

The SLS energy levels are determined by two mechanisms for fixed alloy compositions in the layers. First, the superlattice band energies relative to the bottom of the appropriate quantum wells are determined by the width, height, and separation of the wells. In particular, as a quantum well width is increased with fixed well separation, the minimum energies of the superlattice bands in that well decrease with respect to the bottom of the well [quantum size effect, compare Eq. (3.2.2)]. This is the usual behaviour seen in lattice matched superlattices. Secondly, the energy values of the bottoms and heights of the wells are affected by the strains in the layers, and these strains are determined by the ratio of layer thicknesses.Thus,as a quantum well width is changed with fixed well separation, the energies of the bottom and height of the well change. This behaviour is unique to SLSs.

The calculated results for the lowest conduction subband energy and the uppermost valence subband energy of the $GaAs\text{-}GaAs_{0.2}P_{0.8}$ SLS as a function of the GaAs layer thickness (the well width) for GaAsP layer thickness (the barrier thickness) equal to one lattice constant are shown in Fig. 3.7.a. These energies are given with respect to the unstrained GaAs valence band maximum (taken as the reference level). The relevant quantum wells which determine the energy of the uppermost valence band are associated with the $M_j = \pm 3/2$ hole band minima in the two layers, with the bottoms of the wells occurring in the GaAs layer.As the GaAs layer thicknesses are increased, the quantum well widths also increase so that the superlattice valence band energy approaches the bottoms of the wells. In addition, the changing layer thickness ratio leads to decreasing strain in the GaAs layers and increasing strain in the GaAsP layers. This reduces the valence band splitting in GaAs so that the energy of the bottom of the valence band well shifts, as shown by the lowest dashed line in Fig. 3.7a. The combination of these two effects produces a SLS valence band maximum which increase in energy.

The behaviour of the lowest SLS conduction band states is complicated by the occurrence of two types of bulk band minima at similar energies in the GaAs and GaAsP layers. By examining the tight binding wave function of the lowest energy SLS conduction band state, the relative importance of these various minima can be determined. The non-zero squared amplitudes of the atomic orbitals of the conduction band minimum state for a SLS $d_{GaAs} = 2a_0$ and $d_{GaAsP} = a_0$ indicate |84| that the state character is essentially made up of cation s-orbitals and anion s-orbitals in both layers. This combination of orbitals is characteristic of states associated with the Γ minima in both materials (Bloch states for GaAs, decaying evanescent states for GaAsP). Thus, the quantum wells for

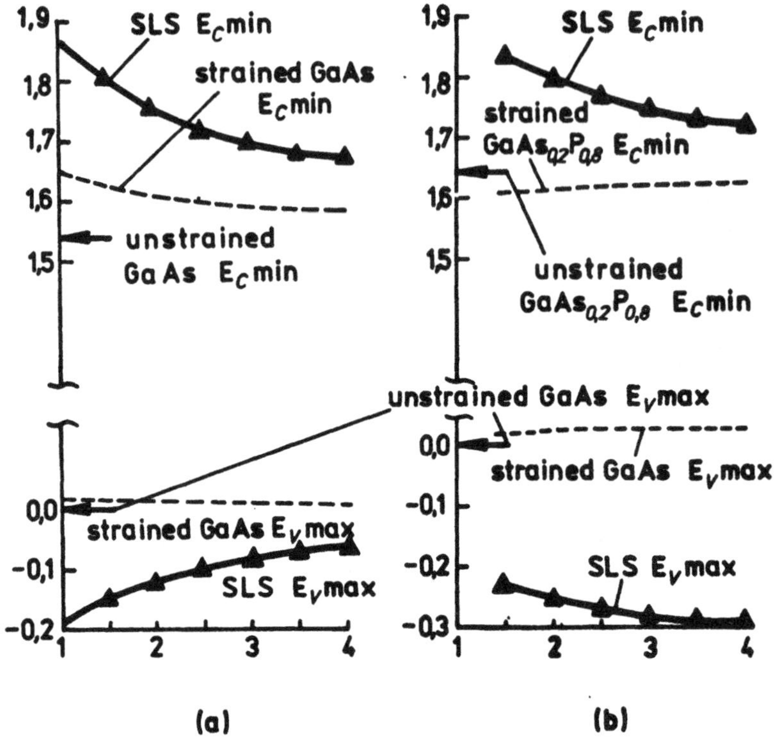

Fig.3.7. Lowest SLS conduction band energy and uppermost SLS valence band energy as a function: a) of GaAs layer thickness (in units of lattice constant) for fixed $GaAs_{0.2}P_{0.}$ layer thickness equal to one lattice constant, and b) of $GaAs_{0.2}P_{0.8}$ layer thickness for fixed GaAs layer thickness equal to one lattice constant. The dashed lines indicate the strained GaAs bulk bands extrema in (a), and the strained $GaAs_{0.2}P_{0.8}$ bulk conduction band minimum and the strained GaAs bulk valence band maximum in (b). For comparison the energy positions of the relevant band extrema for unstrained bulk materials are indicated by arrows (after Osbourn,Ref.84).

this case are determined by the lowest GaAs Γ minimum and the higher lying Γ minimum in GaAsP, even though the GaAsP minima at X are closer in energy to the GaAs Γ minimum. This behaviour results from the poor coupling of GaAs Γ states to GaAsP X states compared to the Γ-Γ state coupling, and is similar to results seen in theoretical studies of the GaAs-AlGaAs system. For the results in Fig. 3.7a, where $d_{GaAs} > d_{GaAsP}$ the Γ type quantum wells determine the SLS conduction band minima energies. The bottoms of these wells are given by the upper dashed line in this figure. As d_{GaAs} is increased, the quantum wells widen so that the lowest superlattice energy drops toward the bottom of the well. Again, the strains in the GaAs layers decrease, so that the bottoms of the wells also drop in energy. The combination of these effects produces a lowering of the SLS conduction band minimum (upper solid line) with increasing d_{GaAs}.

Results for the lowest conduction band energy and the uppermost valence band energy of GaAs/GaAsP as a function of d_{GaAsP} for fixed $d_{GaAs} = a_0$ are given in Fig. 3.7b. Again the energies are given with respect to the unstrained GaAs valence band maximum. The quantum well bottom for the valence band states is given by the lower dashed line in this figure and shifts as the strain in the GaAs layers increases. Increasing d_{GaAsP} corresponds to increasing the separation of the valence band wells, which causes the superlattice state to move up in energy from the well bottom. The combination of these effects leads to a decrease in the SLS valence band maximum energy with increasing d_{GaAsP}.

The lowest SLS conduction band energy drops in energy with increasing d_{GaAsP}, contrary to the behaviour expected for states determined by Γ type quantum wells. The wave functions for the conduction band minimum of a SLS with $d_{GaAs} = a_0$ and $d_{GaAsP} = 2a_0$ indicate, that the state is essentially made up of anion s-orbitals and cation p_x-orbitals in both layers. This combination of orbitals is characteristic of states associated with the (100) X minima in both materials. Thus, the quantum wells for this case are determined by the lowest GaAsP (100) X minimum and the higher lying (100) X minimum in GaAs. The compressive (100) uniaxial strain in the GaAsP layers splits the bulk (010) and (001) conduction band minima to higher energy and the (100) minima to lower energy. For the situation when $d_{GaAsP} > d_{GaAs}$, the (100) X type quantum wells determine the SLS conduction band energies.

In general, the conduction band minimum of an arbitrary GaAs/$GaAs_{0.2}P_{0.8}$ SLS will be the lowest of the two types of states associated with the bulk Γ and bulk X type quantum wells. Since the bulk Γ type states of the layers map onto the Γ point of the SLS Brillouin zone, all of the SLSs in Fig. 3.7a have direct gaps The bulk (100) type states of the layers map onto: (i) the SLS Γ point for structures with an even number of GaAs+GaAsP monolayers (M) per unit cell, and (ii) the zone edge point $2\pi/a_0$ (1/M,0,0) for structures with odd M. Thus, half of the GaAs/GaAsP SLS structures in Fig. 3.7b have direct gaps.

Concluding these considerations, it should be emphasized that the superlattice energies depend on the layer thicknesses through quantum mechanical size effects and through the strains in the layers. For the $GaAs/GaAs_{0.2}P_{0.8}$ (100) SLS, the conduction band minimum is one of two types of states associated with independent Γ and X type quantum wells. Most of these SLSs are found to be direct gap materials.
Another material system creating SLSs with interesting electronic properties is the $In_xGa_{1-x}As$-$In_yGa_{1-y}As$ system. Theoretical calculations have evidenced |88|,that in these all-ternary-compound SLSs the lattice constant, the band gap, and the electron transport properties (for transport in the direction perpendicular to the superlattice interfaces) can be independently varied. This new capability permits the InGaAs SLSs to significantly broaden the range of applications associated with this ternary compound.

The opposed uniaxial strains in the two layer materials of a SLS lead to a system band gap which can be varied by control of the lattice mismatch and of the layer thickness, which allows significantly more flexibility for design of semiconductor devices.

The applicability of SLS systems to devices is limited by an instability of lattice registry which develops at larger mismatches Although maintenance of registry in a very low defect SLS has been limited, as a practical matter, to systems possessing 1-2% mismatch, the mechanism leading to the loss of registry has not been identified.
Stability of lattice registry as a function of lattice parameter mismatch has been studied in |89| for a single interface. The interface model consists of two two-dimensional lattice systems each 12 atoms high by 120 atoms long. Each of the two lattice systems has a different lattice parameter, but is initially constrained to match at the interface. The atoms interact with their nearest neighbours either through a Lennard-Jones potential or an

anisotropic potential designed to more closely simulate covalent bonding. The stability of registry has then been determined via Monte Carlo evolution of the initial state of the system. It has been found that registry is stable to extremely large mismatches (>15%), from which one can infer that the observed loss of registry above a few percent mismatch is not due to instability of an initially perfect superlattice, but rather to an inability to grow perfect interfaces.

The theoretical study of the electron and hole effective masses along the layers of (100) SLSs grown from direct and indirect gap materials has been presented in |90|. It has been shown that the large layer strains in appropriately chosen structures can be used to significantly reduce the 2DEG and 2DHG effective masses associated with zone folded X valleys below the corresponding values in unstrained bulk materials. It is obvious that these results may have important implications for the low field transport properties of strained-layer systems, however no more detailed data are available at present.

3.3. Energy subbands in type II superlattices

From the theoretical point of view, the type II superlattices can be distinguished from that of type I, by the relative locations of the bandedge energies in the two constituent semiconductors (see,Fig.1.1). The fact that in type II superlattices the conduction band of one material is close to the valence band of the other, causes strong band interactions to occur. Whereas in type I superlattices both the electron and the hole states are primarily confined in the narrow-gap semiconductor regions, in type II superlattices the electron states and the hole states are spatially separated (see,Fig.1.3). In the most representative type II superlattice material system, i.e., in the InAs-GaSb system, the top of the GaSb valence band lies higher in energy than the bottom of the InAs conduction band. Therefore, charge transfer across each interface of the periodic array and special electronic properties are to be expected. Experiments indeed, indicate a transition from a semiconductor to a semimetal occurring as the superlattice layer thicknesses are increased, and the lowest conduction-band-like subband sinks through the highest heavy-hole-like subband. The nature of this transition depends strongly on the superlattice

band structure.
It is, in fact, well known that in a general low-symmetry direction in the $\vec{k}$ space, no band degeneracies are allowed, so that a gap must appear between the two intersecting bands |20|. The occurrence of the semimetallic state depends, therefore, on the possibility of crossing along high-symmetry directions. A further mechanism takes place if an indirect gap becomes negative, so that pockets of electrons and holes at different points of the Brillouin zone result. This cannot be determined on symmetry grounds, but must be investigated by detailed calculations.

It has been shown |20| that neither mechanism is operative in the InAs-GaSb superlattice, so that in an intrinsic sample no truly semimetallic band structure results, but rather a zero-gap or very small gap ($E_g \leqslant 10$ meV) semiconductor. Therefore, doping and temperature effects must play a key role in the observed semimetallic behaviour.
There are some stringent requirements on a band-structure calculations for type II superlattices. They will be illustrated on the example of the InAs-GaSb superlattice. In fact, (i) the narrow gap nature of InAs and GaSb, and the mixing of valence and conduction bands at the interfaces, make a many-band treatment necessary, (ii) the investigation of the bands for $\vec{k}$ parallel to the layers requires a realistic description of the host bands, and in particular, of the degenerate valence band-edge, (iii) the occurrence of charge transfer requires a method suitable for self-consistent calculations. The most convenient framework for this problem appears to be the envelope-function approximation, as pointed out by Sham and co-workers |16| and by Bastard |17|,|18|. It has also been shown |20| that besides fulfilling requirement (i) as pointed out by these authors, this method can be formulated so as to fulfil requirements (ii) and (iii) as well.

3.3.1. The envelope-function approximation

Consider an A-B superlattice composed of layers of thickness d_A and d_B, i.e., with period $d = d_A+d_B$. Let z denote the superlattice axis and let x,y denote two orthogonal directions in the z = 0 plane. Following Ref.17 we assume that the band structures of host materials in the vicinity of the centre of the Brillouin zone are quite similar and are well described by the Kane model |91|. For a given superlattice A-B the wave function in each host material takes the form

$$\psi_{A,B}(\vec{r}) = \sum_j F_j^{A,B}(\vec{r}) u_{j0}^{A,B}(\vec{r}) \tag{3.3.1}$$

where $F_j(\vec{r})$ are slowly varying envelope functions and u_{j0} is the periodic part of the Bloch functions at k = 0. If, for simplicity (but this is not crucial), we neglect the Γ_7 spin-orbit split-off valence band, the $F_j(\vec{r})$ are the solutions of a 6 x 6 differential system

$$\sum_{j'} D_{jj'} F_{j'}(\vec{r}) = \lambda F_j(\vec{r}) \tag{3.3.2}$$

where the explicit form of $D_{jj'}$ has been given by Kane and λ is the energy. Equivalently, the eigenvalue problem can be expressed in terms of a nonlinear-in-λ 2 x 2 Hamiltonian. This Hamiltonian acts on the envelope functions associated with two s-type band-edge Bloch functions [hereafter labeled as (1,2)]. Assuming the Kane matrix element π to be identical in A and B host materials, and neglecting the free-electron kinetic energy $p^2/(2m_0)$ one finds

$$\begin{bmatrix} H_{11} & H_{12} \\ H_{21} & H_{22} \end{bmatrix} \begin{bmatrix} F_1 \\ F_2 \end{bmatrix} = \lambda \begin{bmatrix} F_1 \\ F_2 \end{bmatrix} . \tag{3.3.3}$$

where

$$H_{11} = H_{22} = V_s + \pi^2 p_- \frac{1}{E_{gA}+\lambda-V_p} p_+ + (\pi^2/3) p_+ \frac{1}{E_{gA}+\lambda-V_p} p_- +$$

$$+ (2\pi^2/3) p_z \frac{1}{E_{gA}+\lambda-V_p} p_z \tag{3.3.4}$$

$$H_{12} = H_{21}^* = (\pi^2\sqrt{2/3}) \left[p_z \frac{1}{E_{gA}+\lambda-V_p} p_+ - p_+ \frac{1}{E_{gA}+\lambda-V_p} p_z \right]$$

where $p_\pm = (p_x \pm ip_y)/\sqrt{2}$. The energy origin has been taken at the bottom of the A conduction band, and E_{gA} is the energy gap in the A material. V_s and V_p are the shifts of the s and p levels at the Γ point when going from A to B host materials.

We recall that Eq. (3.3.3) and (3.3.4) are valid in each host material but do not give any information on the potential energy associated with the A-B interfaces. Actually, for infinite A or B layers, the (1,2) coupling disappears, and the eigenenergies of (3.3.3) are the well-known Kane dispersion relations. For instance, in material A

$(2\pi^2)/(3\hbar^2)(k_z^2 + 2k_+k_-) = \lambda(\lambda + E_{gA})$. We will assume the interface potential to be strongly localized in the vicinity of the A-B interfaces. Its spatial localization is such that at the scale of variation of the varying envelopes the interfaces reduce to the planes

$$\left.\begin{array}{l} z = nd \\ z = d_A + nd \end{array}\right\} \text{ n are integers} \qquad (3.3.5)$$

Moreover, we will assume the interface potential does not mix but only shifts the s- and p-like states at the band-edges. For a perfect interface this seems plausible, since A and B have the same spatial symmetry and the same lattice constants. With these "Ansätze" the only effect of the interface on the envelope functions F_1 and F_2 is to make the V_s and V_p appearing in Eqs. (3.3.4) depend on z through

$$V_{s(p)}(z+d) = V_{s(p)}(z)$$

and

$$V_{s(p)}(z) = \begin{cases} Q \text{ if } 0<z<d_A \\ V_{s(p)} \text{ if } d_A<z<d_A+d_B = d \end{cases} \qquad (3.3.6)$$

The z dependence although quite simple, has important consequences. Namely, the (1,2) coupling does not vanish in the superlattice since $E_{gA} + \lambda - V_p$ does not commute with p_z. In turn, this coupling influences the boundary conditions which F_1 and F_2 should fulfil at the A-B interfaces.

We will assume the envelope functions F_1, F_2 to be continuous at the interface. To obtain two other boundary conditions, we integrate Eqs. (3.3.3) and (3.3.4) across an interface. Since the 2x2 Hamiltonian [Eq. (3.3.3)] is independent of (x,y), $\vec{k}_t = (k_x, k_y)$ is a good quantum number. As long as $V_s(z)$ and $V_p(z)$ have at most finite jumps and making use of the F_1, F_2 continuity, we obtain

$$[E_{GA} + \lambda - V_p(z)]^{-1}(-\sqrt{2}\, F_1' - ik_+F_2)$$

and (3.3.7)

$$[E_{gA} + \lambda - V_p(z)]^{-1}(-\sqrt{2}\, F_2' + ik_-F_1)$$

both expressions continuous at the interface, where the prime denotes the derivative with respect to z, and $k_{\pm} = (k_x \pm ik_y)/\sqrt{2}$. Furthermore, the long-range behaviour of $F_{1,2}$ is governed by the Bloch theorem,

$$F_{1,2}(z+md) = \exp(iqmd)F_{1,2}(z) \qquad (3.3.8)$$

valid for any z and any relative integer m. In Eq. (3.3.8) q is the superlattice wave vector along the z-axis.

So far the considered theory is general enough to be applicable to type I, as well as to type II superlattices. In the case of the type I superlattices it leads to a dispersion relation similar to the Eq. (3.2.32). In the case of the type II InAs-GaSb superlattice, however, two important facts introduce noticable differences. First, the signs of the effective masses are for the host materials different ($m_A^*>0$, $m_B^*<0$). Secondly, there exist a symmetry mismatch between the states of each host material, which have to be admixed in the superlattice state (the conduction states in InAs are predominantly s-like, whereas light-hole GaSb levels are predominantly p -like. Moreover, this symmetry mismatch is preserved across the interface. The building of superlattice states in this case is possible only because of p-admixture in the InAs electron states and of s-admixture in the GaSb light-hole levels.

Using Eqs. (3.3.7) and (3.3.8) for the derivation of the dispersion relation relevant to the InAs-GaSb superlattice, and taking the energy reference level (the zero of energy) at the bottom of the InAs conduction band, one obtains

$$\left.\begin{array}{l} V_p = \Delta + E_g(\text{InAs}) = 0.56 \text{ eV} \\ V_s = \Delta + E_g(\text{GaSb}) = 0.96 \text{ eV} \end{array}\right\} \text{if z corresponds to a GaSb layer}$$
$$V_s = V_p = 0 \quad \text{if z corresponds to an InAs layer} \qquad (3.3.9)$$

The overlap parameter Δ between GaSb valence band and the InAs conduction band has been taken to be equal to 0.15 eV |92|. Let us assume for simplicity $\vec{k}_t = 0$. Then the (1,2) mixing disappears. The superlattice states originating from the matching of InAs conduction states with GaSb light-hole levels have dispersion relations given by the following equation

$$\cos(qd) = \cos(k_A d_A)\cos(k_B d_B) - (1/2)(x+x^{-1})\sin(k_A d_A)\sin(k_B d_B) \qquad (3.3.10)$$

where

$$(2\pi^2/3)k_A^2 = \lambda[\lambda + E_g(\mathrm{InAs})] \tag{3.3.11}$$

$$(2\pi^2/3)k_B^2 = (\lambda-\Delta)[\lambda-\Delta-E_g(\mathrm{GaSb})] \tag{3.3.12}$$

$$x = (k_A/k_B)(\lambda-\Delta)/[\lambda+E_g(\mathrm{InAs})] \tag{3.3.13}$$

In the parabolic limit, for $\lambda \ll E_q(\mathrm{InAs})$, one obtains

$$x \approx x_{parab.} = (k_A/k_B)(m_B^*/m_A^*)(\Delta-\lambda)/[E_g(\mathrm{GaSb})] \tag{3.3.14}$$

Note that in Eq. (3.3.14) m_B^* is negative.

Assuming $d_A = d_B = d/2$, the evolution with increasing periodicity d of the superlattice band structure, which arises from the mixing of InAs electron states with GaSb light-hole levels, can be calculated according to Eqs. (3.3.10)-(3.3.13). The results are shown in Fig. 3.8.

Equations (3.3.10)-(3.3.13) do not apply to the superlattice bands which originate from the heavy-hole bands of the host materials. This arises from the fact that heavy-hole states correspond to vanishing $F_{1,2}$ envelope functions, i.e., the heavy-hole curvature is governed by remote bands not included to the simplified Kane model used throughout these considerations.

Still one may accurately calculate the heavy-hole superlattice bands if $\vec{k}_t = 0$. The effect of remote bands in each host material can be incorporated up to order k^2 in the 4 x 4 matrix acting on the F_i, i = 3-6 envelope functions. If $\vec{k}_t = 0$, the modified 4 x 4 matrix is diagonal and the heavy-hole states correspond to $J = 3/2$, $M_J = \pm 3/2$, i.e., they are entirely decoupled from light--hole states $M_J = \pm 1/2$. The $M_J = \pm 3/2$ modified matrix elements are simply

$$D_{3/2,-3/2} = -p_z^2/[2m_{hh}^*(A,B)] \tag{3.3.15}$$

where $m_{hh}^*(A,B)$ are the heavy-hole effective masses in materials A and B. Owing to the complete decoupling between heavy holes on the one hand and electrons or light holes on the other, one can see that the heavy-hole superlattice bands can be calculated exactly in the same way as superlattice conduction states in "parabolic" type I superlattices. In the InAs-GaSb material system,

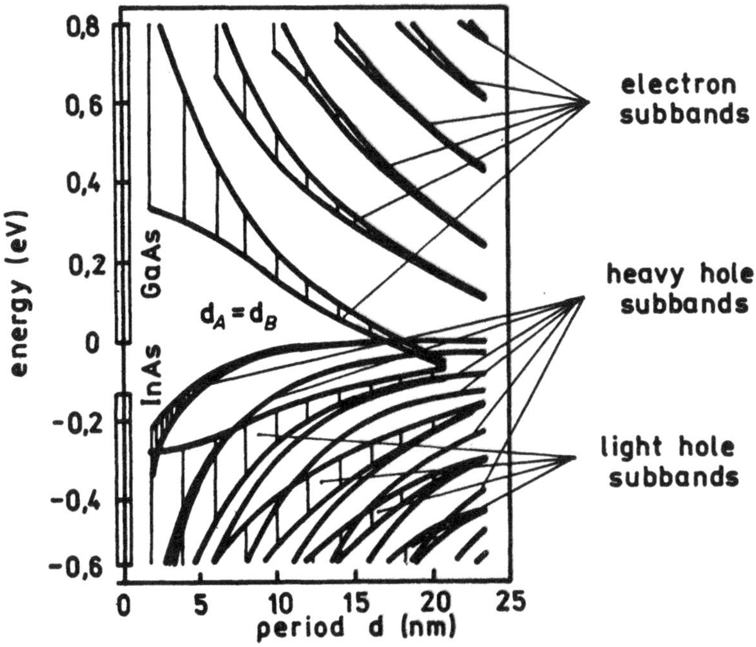

Fig.3.8. Calculated subband energies and widths for electrons and heavy and light holes as a function of the period thickness $d(d_A = d_B)$ in the InAs-GaSb superlattice. The cross--over of the electron and heavy hole subbands at d=17 nm indicates the semiconductor-semimetal transition (after Chang et al., Ref.92).

the GaSb layers are potential wells for heavy holes. Measuring the energy from the top of the GaSb (B material here) valence band, and denoting by λ_{hh} the heavy-hole energy, the dispersion relations $\lambda_{hh}(q)$ are given by

$$\cos(qd) = \cos(k_A d_A)\cos(k_B d_B) - (1/2)(x + x^{-1})\sin(k_A d_A)\sin(k_B d_B) \tag{3.3.16}$$

where

$$k_A = \{[2m^*_{hh}(A)/\hbar^2](\lambda_{hh} - V_{hh})\}^{1/2} \tag{3.3.17}$$

$$k_B = \{[2m^*_{hh}(B)/\hbar^2]\lambda_{hh}\}^{1/2}$$

where $m^*_{hh}(A) = 0.4\ m_0$, $m^*_{hh}(B) = 0.33\ m_0$, and $V_{hh} = 0.56$ eV. Since V_{hh} is quite large, the superlattice heavy-hole bands are almost dispersionless, except for small ($d_A < 2$ nm) InAs thicknesses.

Numerical calculations performed for $d_A = d_B$ indicate (Fig.3.8) that the ground heavy-hole band behaves like the ground bound level of a particle confined in a quantum well of infinite depth if $d_B > 15$ nm.

In conclusion, it should be emphasized that the advantage of the envelope-function formalism lies in the simplicity of the final results, Eqs. (3.3.10)-(3.3.13), as well as in their generality. Both kinds of the existing compositional superlattices are described by the same formalism and the differences in their band structures are naturally explained. Finally, the envelope-function approximation is able to incorporate finite $\vec{k}_t$. This will permit a discussion of the transverse motion (i.e., in the layer plane), which is in fact most frequently involved in the experiments.

3.3.2. The reduced Hamiltonian method for solving the superlattice tight-binding equations

Theoretical calculations of superlattice electronic properties commonly use some of the bulk information as inputs |15|. For example, the superlattice tight-binding calculations use empirical matrix elements determined by fitting to bulk band structures |8| Similarly, the various envelope-function methods use bulk effective masses, band gaps, and momentum matrix elements |16|,|18|,|20|.

Despite the use of the bulk information, the results of the superlattice calculations- energy bands and wave functions - are not always easily interpreted in terms of the bulk properties. The concept of zone folding represents perhaps the simplest point of view relating superlattice and bulk band structures. It provides little quantitative information, however.

The envelope-function approximation method is a useful method that is simple enough so that the relationship between the superlattice and bulk wave functions is easily extracted. There are important situations, however, where a more complete description than can be provided by the envelope-function approximation is necessary to obtain an understanding of the superlattice electronic properties. This can occur when two or more bulk states with widely separated wave vectors contribute significantly to a given superlattice state.

In an attempt to explore more fully the relationship between superlattice and bulk electronic properties, the empirical tight-binding model has been used |15|. It has the advantage of incorporating several bulk bands (20 including spin) simultaneously over the full Brillouin zone. The tight-binding model describes the wave function, bulk or superlattice, as a linear combination of atomic-like orbitals. In order to solve the tight-binding equations for the superlattice the reduced Hamiltonian method |13|, |14| has been applied. This method has three major components: transfer matrices, an expansion in bulk states with complex wave-vector values, and a quickly convergent iterative method for finding energy eigenvalues. The transfer matrices relate the electronic wave function on neighbouring atomic planes in non-bulk-like regions, i.e., regions in which there is interfacial disruption. Sufficiently far from the interface, the tight-binding Hamiltonian is bulk-like and, therefore, the wave function can be expressed as a linear combination of bulk states. In addition to the familiar oscillating Bloch bulk states, the presence of the interface requires that bulk states, which grow or decay exponentially away from the interface, be included in the expansion. These are bulk states which have complex wave-vector values.

The transfer matrices and complex wave-vector bulk states are then used in the formation of a "reduced" tight-binding Hamiltonian matrix for the system. Diagonalizing it produces the energy eigenvalues and the electronic wave function. However, the transfer matrices and bulk-state expansion are found assuming an initial guess for the energy. Thus, this method must be iterated until the assumed energy agrees with a final energy eigenvalue of the Hamiltonian matrix. Convergence occurs rapidly, typically within five iterations for milielectron volt precision.

The reduced Hamiltonian method has also been applied for studying the band mixing effects in superlattices |15|. Let us consider for example the InAs-GaSb (001) superlattice. The bulk band structure of InAs and GaSb are assumed to be known. Except for smaller effective masses, the band structures are similar to that of GaAs. The conduction band minimum of InAs lies slightly below the valence band maximum of GaSb. There have been various attempts to determine the magnitude of the energy difference between the two levels, Δ both by using empirical rules and by fitting to experiment. The values of Δ differ between 0.100 eV |15| and 0.175 eV |16|.

Since the superlattice conduction band minimum is derived from the InAs conduction band-edge and its valence band maximum from the GaSb heavy-hole band, the superlattice band gaps are quite sensitive to the InAs conduction band minimum and GaSb heavy-hole band maximum effective masses assumed. Much of the variation in Δ reported can, therefore, be attributed to the variability of the assumed values of the effective masses, and not just to the difference between theoretical models. Assuming Δ = 0.1 eV the following results have been obtained |15|:

(i) most of the InAs-GaSb superlattice states are clearly identifiable as conduction, heavy-hole, or light-hole-like in character. The two, highest in energy, Brillouin-zone-centre states, are conduction band states, with their amplitudes mostly on the InAs layers. The six lower-energy states are all in the valence band and are mostly on the GaSb layers.

(ii) The two conduction band states do have significant contributions from the X-derived bulk components in the InAs and GaSb layers, especially the higher state. Due to the relatively large magnitudes of the imaginary parts of the X-derived wave vectors in the energy range of interest, their influence on the superlattice wave functions are confined near the interfaces to within one atomic layer. Due to them, however, the wave function in the InAs layers does not match smoothly to that in the GaSb layers. Instead, there is a noticeable discontinuity in the tight-binding envelope function at the interface. Conventional envelope-function methods are not capable of duplicating this feature because the bulk band structure near the X point is omitted. Thus, the boundary conditions at the interfaces assumed by the envelope-function method may not be accurate in certain cases.

(iii) The valence band states are clearly identifiable as either heavy-hole-like or light-hole-like, particularly within the GaSb (well) layers. There are four heavy-hole states and two light-hole states. The states are, however, not pure. The lowest-in-energy, mainly light-hole superlattice state in particular has a large bulk heavy-hole component, especially in the InAs layers. The spin-orbit split-off bulk state also has a not insignificant contribution to several of the valence band states, even though the bulk split-off band maximum is several tenths of electron volts lower in energy. The X-point-derived bulk components have small contributions to the valence band states. Again, the bulk components with wave vectors having nonzero imaginary parts contribute mainly near the interfaces. Since the envelope-function approximation does not take into account this mixing of bulk components for superlattice Brillouin zone-centre wave functions, its treatment of the boundary conditions may be inadequate here.

In conclusion, one can realize that the electronic states of semiconductor superlattices can be conveniently expressed as linear combinations of bulk states of the constituent semiconductors. In most cases, the Γ point superlattices states were identifiable as being predominantly composed of one type of bulk state. Admixtures of other bulk states served to modify the energies of the states somewhat and to cause wave function discontinuities at the interfaces as compared with the envelope-function approximation. More mixing was found for the InAs-GaSb superlattice as compared to the GaAs-AlAs superlattice due to the greater bulk potential differences of its constituents. For certain alloy compositions and layer thicknesses, especially large mixing of heavy- and light-hole components occur in the second and third valence subbands.

3.4. Polytype superlattice band structure

Our considerations have been so far restricted to binary A-B superlattices. It may prove useful for specific technological purposes to grow ternary A-B-C, quaternary A-B-C-D, etc., superlattices. The polytype superlattices described in Sec. 1.2 may serve here as an example. The calculations of A-B-C type superlattice band structure will be almost unfeasible in the three-dimensional LCAO (linear combination of atomic orbitals) framework unless d_A, d_B, d_C are only a few atomic planes thick. On the other hand, the envelope-function approximation is capable of achieving this task.

To do so, let us use the transfer matrix scheme |18|. Consider a given ABC superlattice unit cell and assume for simplicity $\vec{k}_t=0$. Inside each layer the projection $F_s(z)$ of the wave function on the s level is a sum of incoming and outgoing plane waves:

$$F_s(z) = A_A = \exp(ik_A z) + B_A \exp(-ik_A z), \quad 0 \leq z \leq d_A$$
$$F_s(z) = A_B \exp[ik_B(z-d_A)] + B_B \exp[-ik_B(z-d_A)], \quad d_A \leq z \leq d_A+d_B \qquad (3.4.1)$$
$$F_s(z) = A_C \exp[ik_C(z-d_A-d_B)] + B_C \exp[-ik_C(z-d_A-d_B)], \quad d_A+d_B \leq z \leq d_A+d_B+d_C$$

The boundary conditions for F_s and its derivative at the A-B and B-C interfaces can be written in the form

$$\begin{bmatrix} A_B \\ \\ B_B \end{bmatrix} = T_{A\to B} \begin{bmatrix} A_A \\ \\ B_A \end{bmatrix}, \quad \begin{bmatrix} A_C \\ \\ B_C \end{bmatrix} = T_{B\to C} \begin{bmatrix} A_B \\ \\ B_B \end{bmatrix} \qquad (3.4.2)$$

and the Bloch theorem reads

$$\begin{bmatrix} A_A \\ \\ B_A \end{bmatrix} \exp(iqd) = T_{C\to A} \begin{bmatrix} A_C \\ \\ B_C \end{bmatrix} \qquad (3.4.3)$$

(A_A, B_A) is then the eigenvector corresponding to the eigenvalue $\exp(iqd)$ of the transfer matrix

$$t = T_{C\to A} T_{B\to C} T_{A\to B} \qquad (3.4.4)$$

To find the eigenvalue of t the det t has to be 1 and the solutions are such that |11|

$$2\cos(qd) = \mathrm{Tr}(t) \qquad (3.4.5)$$

Let us choose the energy zero at the bottom of the s-type band of the A material. Define $V_p^{B,A}$ and $V_p^{C,A}$ the energy shifts of the p-edges between A and B,A and C materials, respectively. In terms of the quantities

$$x_{A,B} = (k_A/k_B)(E_{gA} + \lambda - V_p^{A,B})(E_{gA} + \lambda)^{-1} \qquad (3.4.6)$$
$$x_{B,C} = (k_B/k_C)(E_{gA} + \lambda - V_p^{C,A})(E_{gA} + \lambda - V_p^{B,A})^{-1}$$
$$x_{C,A} = (k_C/k_A)(E_{gA} + \lambda)(E_{gA} + \lambda - V_p^{C,A})^{-1}$$

which are such that $x_{A,B} x_{B,C} x_{C,A} = 1$, the dispersion relations obtained from Eq. (3.4.5) are

$$\cos(qd) = \cos(k_A d_A)\cos(k_B d_B)\cos(k_C d_C) - (1/2)[X_{B,C}\cos(k_A d_A)\cdot$$
$$\cdot\sin(k_B d_B)\sin(k_C d_C) + X_{C,A}\sin(k_A d_A)\cos(k_B d_B)\sin(k_C d_C) +$$
$$+ X_{A,B}\sin(k_A d_A)\sin(k_B d_B)\cos(k_C d_C) \qquad (3.4.7)$$

with

$$X_{i,j} = x_{i,j} + x_{i,j}^{-1} \qquad (3.4.8)$$

where q is the superlattice wave vector and $d = d_A + d_B + d_C$, the superlattice period. This procedure can be clearly iterated. For instance, the dispersion relations of ABCD-type superlattice are obtained in terms of $x_{A,B}$, $x_{B,C}$, $x_{C,D}$, $x_{D,A}$:

$$
\begin{aligned}
\cos(qd) = {} & \cos(k_A d_A)\big[\cos(k_B d_B)\cos(k_C d_C)\cos(k_D d_D) - (1/2)(x_{C,D}x_{B,C} + \\
& + x_{C,D}^{-1}\, x_{B,C}^{-1})\sin(k_B d_B)\cos(k_C d_C)\sin(k_D d_D) - (1/2)X_{C,D}\,\sin(k_C d_C) \\
& \sin(k_D d_D)\cos(k_B d_B) - (1/2)X_{B,C}\,\sin(k_B d_B)\sin(k_C d_C)\cos(k_D d_D)\big] - \\
& - (1/2)\sin(k_A d_A)\{(x_{A,B}x_{B,C} + x_{A,B}^{-1}x_{B,C}^{-1})\cos(k_D d_D)\cos(k_B d_B)\sin(k_C d_C) - \\
& - (x_{B,C}x_{D,A} + x_{B,C}^{-1}x_{D,A}^{-1})\sin(k_B d_B)\sin(k_C d_C)\sin(k_D d_D) + \\
& + \cos(k_C d_C)\big[X_{A,B}\,\sin(k_B d_B)\cos(k_D d_D) + X_{D,A}\,\sin(k_D d_D)\cos(k_B d_B)\big]\}
\end{aligned}
\tag{3.4.9}
$$

where all the symbols have the same meaning as before except that d is now equal to $d_A+d_B+d_C+d_D$.

Let us remark that for the ABC system there exist in principle two possible ternary superlattices built from these elements. There are the ABC and ACB materials. However, the sequence ACBACB... is ABCABC...spelled backward. Therefore, the superlattice potentials $V_{ACB}(z)$ and $V_{ABC}(z)$ are such that $V_{ACB}(z) = V_{ABC}(-z)$. The Bloch eigenstates $\psi_{ABC}(z)$ and $\psi_{ACB}(z)$ of the same energy correspond to superlattice states of opposite wave vector q and -q. But $\lambda(q) = \lambda(-q)$ and the dispersion relations of ABC and ACB superlattices are, therefore, identical. This is in fact apparent in Eq. (3.4.7) which is invariant under any permutation of A,B, and C. For quaternary superlattices such a simplification does not occur and the ABCD,ACBD superlattices should have different band structures.

3.5. Electronic structure of doping superlattices

A doping superlattice crystal (a nipi-crystal) consists of an arbitrary homogeneous semiconductor with a periodic variation of n- and p-type doping

$$N_D(z+d) = N_D(z) \quad \text{and} \quad N_A(z+d) = N_A(z) \tag{3.5.1}$$

where z is the direction of periodicity and d the superlattice period. The ground state of the nipi-crystal can be described quite easily, if there are no other impurities present than the shallow donors and acceptors, if the crystal is "macroscopically compensated", i.e.,

$$\int_{-d/2}^{d/2} N_D(z)\,dz = \int_{-d/2}^{d/2} N_A(z)dz \qquad (3.5.2)$$

and if the superlattice period and the doping concentration do not exceed some limiting values to be discussed below. Under these conditions all the impurities will be ionized and the superlattice potential will be the space-charge potential of the impurity distribution

$$Q(z) = e[N_D(z) - N_A(z)] \qquad (3.5.3)$$

For the sake of simplicity we will restrict ourselves to the case of homogeneous doping as shown in Fig. 3.9, with rectangular doping profiles which are symmetric with respect to the origin placed in the middle of an n-type doped layer.

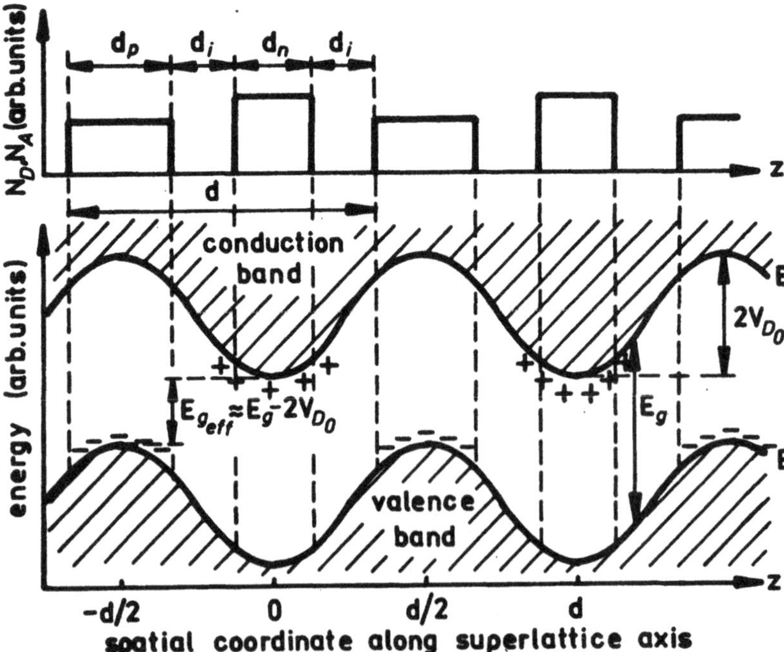

Fig.3.9. Schematic illustration of the properties characteristic for a nipi-crystal with constant impurity concentration in the n- and p-type layers. Plus signs indicate ionized donor levels near the conduction band-edge, and minus signs indicate the negatively charged acceptor levels above the valence band-edges (after Ruden et al.,Ref. 22)

The space-charge potential of the impurities V(z) is obtained by integrating Poisson's equation

$$\frac{\partial^2 V\, z}{\partial z^2} = 4\pi e Q(z)/\varepsilon \qquad (3.5.4)$$

subject to the boundary conditions

$$\left.\frac{\partial V(z)}{\partial z}\right|_{z=0}, \quad V(0) = 0 \qquad (3.5.5)$$

where ε is the static dielectric constant of the semiconductor. V(z) consists of parabolic parts in the doping layers

$$V(z) = \begin{cases} (2\pi e^2 N_D/\varepsilon)z^2 & \text{for } |z| \leq d_n/2 \\ 2V_{Do} - (2\pi e^2 N_A/\varepsilon)(d/2-|z|)^2 & \text{for } d/2 - |z| \leq d_p/2 \end{cases} \qquad (3.5.6)$$

and linear parts in the intrinsic regions

$$V(z) = (2\pi e^2 N_D d_n/\varepsilon)(|z| - d_n/4) \text{ for } d_n/2 \leq |z| \leq (d-d_p)/2 \qquad (3.5.7)$$

The maximum height of V(z) is $2V_{Do}$, given by

$$2V_{Do} = (2\pi e^2/\varepsilon)(N_D d_n^2/4 + N_A d_p^2/4 + N_D d_n d_i) \qquad (3.5.8)$$

For $|z| \geq d/2$, V(z) is obtained from a periodic repetition of the expressions (3.5.6)-(3.5.8).

Figure 3.9 shows the important consequences of the superposition of V(z) to the crystal potential. The effective band gap $E_{g_{eff}}$, i.e., the difference between the lowest conduction- and the uppermost valence band states, is lowered by about $2V_{Do}$ compared with the gap of the homogeneous bulk material E_g. The term "indirect gap in real space" which was used in Sec. 1.3, now becomes clear: The lowest conduction band states are shifted by half a superlattice period with respect to the uppermost valence band states.

It should be noted that the value of $2V_{Do}$ may exceed the band gap E_g of the unmodulated semiconductor. Such a situation implies the possibility of conduction subbands with energies lower than the upper valence subband edge, which would correspond to a negative effective band gap $E_{g_{eff}}$ (compare, Fig. 2.15). Actually, under these conditions there will be a finite electron and hole concentration in the n- and p-type layers in the ground state. The compensating space charge of the charge carriers reduces the value of $2V_{Do}$ and changes the shape of the space-charge potential in a self-consistent way. The ground state will then be characterized by electron and hole concentrations $n_o^{(2)}$ and $p_o^{(2)}$ per layer which make the Fermi level equal for both types of carriers (this is the nipi-semimetal).

Another case with a finite carrier concentration in the ground state occurs whenever the nipi-crystal is not compensated, i.e., condition (3.5.2) does not hold. The ground state superlattice potential, evidently, has to be calculated self-consistently in such a situation also.

One of the peculiarities of nipi-crystals is the possibility to induce metastable deviations from thermal equilibrium with different electron and hole quasi-Fermi levels E_{Fn} and E_{Fp}, by varying of the electron and hole concentration in the layers. Thus, the calculation of the electronic structure of a nipi-crystal is not to be restricted to the ground state. The subband energies and wave functions are required as a function of the tunable carrier concentration. In order to solve the Schrödinger equation for a nipi-crystal some simplifying assumptions are usually made in the calculation of the impurity space-charge potential. So far, also in the numerical calculations related to this problem the space charge of the impurities, is treated as homogeneously smeared out in the respective doping regions. In doing so, the following two facts are neglected: the spatial potential fluctuations which result from the random distribution of impurities in the doped regions, and the point-charge character of the impurities in the layers, which may lead to bound impurity states or to the formation of impurity bands. We will restrict our considerations here only to the most popular nipi-crystal material, i.e., to GaAs. In doping superlattices made of this material |22|:
(i) the influence of potential fluctuations is reasonably small,
(ii) donor impurity bands may be neglected, and
(iii) the formation of acceptor impurity bands must be considered within the typical range of values of design parameters of nipi-crystals.

We will now proceed by discussing the electronic subband structure in the n-type layers in the framework of the local-density-functional formalism |23|,|24|. Together with the effective-mass approximation this formalism leads to a Schrödinger-type equation of the form

$$[-(\hbar^2/2m_e^*)\nabla^2+V(z)+V_H(z)+V_{xc}(z)]\psi_{n\vec{k}}(\vec{r}) = E_{n\vec{k}}\psi_{n\vec{k}}(\vec{r}) \qquad (3.5.9)$$

Since the potential does not depend on x or y and is periodic in z with the periodicity d we can write for the wave functions

$$\psi_{n\vec{k}}(\vec{r}) = (1/\sqrt{A})\exp(i\vec{k}_t\vec{r}_t)F_{nk_z}(z) \qquad (3.5.10)$$

where $\vec{k}_t$ is the wave vector parallel to the doping layers. A is the normalization area and $F_{nk_z}(z)$ is the superlattice Bloch function of subband n and wave vector $k_z(-\pi/d \leq k_z \leq \pi/d)$, with the lattice periodic part $u_{\vec{k}=0}(\vec{r})$ neglected. For the eigenvalues $E_{n\vec{k}}$ we thus get

$$E_{n\vec{k}} = E_{nk_z} + \hbar^2k_t^2/(2m_e^*) \qquad (3.5.11)$$

and the Bloch functions $F_{nk_z}(z)$ and the eigenvalues E_{nk_z} are to be determined self-consistently from

$$[-(\hbar^2/2m_e^*)(\frac{\partial^2}{\partial z^2})+V(z)+V_H(z)+V_{xc}(z)]F_{nk_z}(z) = E_{nk_z}F_{nk_z}(z) \qquad (3.5.12)$$

The Hartree contribution of the electrons to the self-consistent potential is given by the solution of the one-dimensional Poisson's equation (3.5.4) subject to the boundary conditions (3.5.5) in which $V_H(z)$is substituted for V(z), and with the electron density

$$n(z) = 2 \sum_{n\vec{k}} |F_{nk_z}(z)|^2 \, n_n^{(2)} \qquad (3.5.13)$$

where $n_n^{(2)}$ is the number of carriers in the n-th subband, which follow from the condition of equal Fermi level in all populated bands |53|. The local exchange and correlation potential $V_{xc}(z)$ is obtained from

$$V_{xc}(z) = E_{xc}[n(z)] + n(z)(\delta E_{xc}/\delta n) \qquad (3.5.14)$$

where $E_{xc}(n)$ is the exchange and correlation energy per electron of a homogeneous electron gas of the local density n. Since the electronic densities of interest in nipi-crystals are usually very high, i.e., the mean distance between two electrons is short compared to the effective-mass Bohr radius $A_D^{EMA} = a_B\varepsilon(m_0/m_e^*)$, we will take, according to |22|

$$- E_{xc} = (0.916/r_s)[e^2/(2\varepsilon a_D^{EMA})] \tag{3.5.15}$$

with

$$r_s = [4\pi/(3n)]^{1/3}/a_D^{EMA} \tag{3.5.16}$$

The inclusion of higher-order correlation terms does not affect the results appreciably. The contribution of the holes in the p-type layers to the self-consistent potential is not written explicitly since it can be incorporated into V(z).

In most cases of interest only the lowest electronic subbands are partially occupied. For these low subbands and typical doping parameters an extreme tight-binding approach is correct, i.e., the subbands are flat in the k_z direction. The Bloch function can be written as

$$F_{nk_z}(z) = (1/\sqrt{N_{SL}}) \sum_m \exp(ik_z dm)\Phi_n(z-md) \tag{3.5.17}$$

where N_{SL} is the number of superlattice periods in the crystal and $\Phi_n(z-md)$ is the wave function of subband n centred at the n--type layer with label m. Obviously, the square modulus of $F_{nk_z}(z)$ is in this case independent of k_z, and the electron density may be written as

$$n(z) = \sum_n n_n^{(2)}|F_{n0}(z)|^2 N_{SL} = \sum_{n,m} n_n^{(2)}|\Phi_n(z-md)|^2 \tag{3.5.18}$$

where the population $n_n^{(2)}$ of the n-th subband is determined by the requirement of equal Fermi levels in occupied subbands

$$E_n + (\hbar^2/2m_e^*)2\pi n_n^{(2)} = E_0 + (\hbar^2/2m_e^*)2\pi n_0^{(2)} \tag{3.5.19}$$

with

$$\sum_{n_{occ}} n_n^{(2)} = n^{(2)} \tag{3.5.20}$$

In order to solve Eq. (3.5.12) self-consistently by direct numerical integration at k_z=0 and $k_z = \pi/d$, the following boundary conditions have to be imposed |22|:

for n = 0,2,4,...

$$\left.\frac{\partial F_{n0}(z)}{\partial z}\right|_{z=0} = \left.\frac{\partial F_{n,\pi/d}(z)}{\partial z}\right|_{z=0} = 0 \tag{3.5.21}$$

$$\left.\frac{\partial F_{n0}(z)}{\partial z}\right|_{z=d/2} = 0, \quad F_{n,\pi/d}(d/2) = 0 \qquad (3.5.22)$$

and for n = 1,3,5,...

$$F_{n0}(0) = F_{n,\pi/d}(0) = 0 \qquad (3.5.23)$$

$$F_{n0}(d/2) = 0, \quad \left.\frac{\partial F_{n,\pi/d}(z)}{\partial z}\right|_{z=0} = 0 \qquad (3.5.24)$$

Some nipi-crystal subbands with a finite k_z dispersion may become occupied in the very highly excited state. It is then usually sufficient to calculate the potential once with the wave functions associated with $k_z=0$ and once with those at $k_z=\pi/d$ and then to average.

To illustrate the variety of electronic structures obtainable with different design parameters of the system, results for two different superlattice structures will be presented in Figs. 3.10 and 3.11. These are calculated for a structure without intrinsic layers, and for an extreme nipi structure with thin doping layers and rather wide intrinsic layers. To facilitate the comparison of these systems one parameter has been kept fixed, namely, the superlattice period d. As the first structure, the sample composed of 40-nm thick GaAs layers with constant doping concentrations $N_D = N_A = 10^{18}$ cm^{-3} without intrinsic layers has been chosen. The solid lines shown in Fig. 3.10a represent the energies for the bottom of the various subbands referred to the position of the hole quasi-Fermi level E_{F_p} as a function of the two-dimensional carrier concentration per layer. It has been assumed that E_{F_p} coincides with the acceptor level in the central part of the p-type layers (acceptor ionization energy E_A = 28 meV for Be acceptors).

It should be noted that the curve $E_{F_n} - E_{F_p}$ versus $n^{(2)}$ (dashed-doted line) does not deviate appreciably from a parabola with a vertex at $n^{(2)} = N_D d_n = 4\cdot10^{12}$ cm^{-2}. This behaviour reflects the parabolic relation between injected carriers and the quasi-Fermi level distance $E_{F_n}-E_{F_p}$ which would be obtained from a simple classical treatment of the problem. At $n^{(2)}= 0$ all the subbands shown are nearly equally spaced with $E_n-E_{n-1} = \hbar[4\pi e^2N_D/(\varepsilon m_e^*)]^{1/2}$ since the bare potential V(z) deviates appreciably from a harmonic potential only at higher energies. With increasing carrier concen-

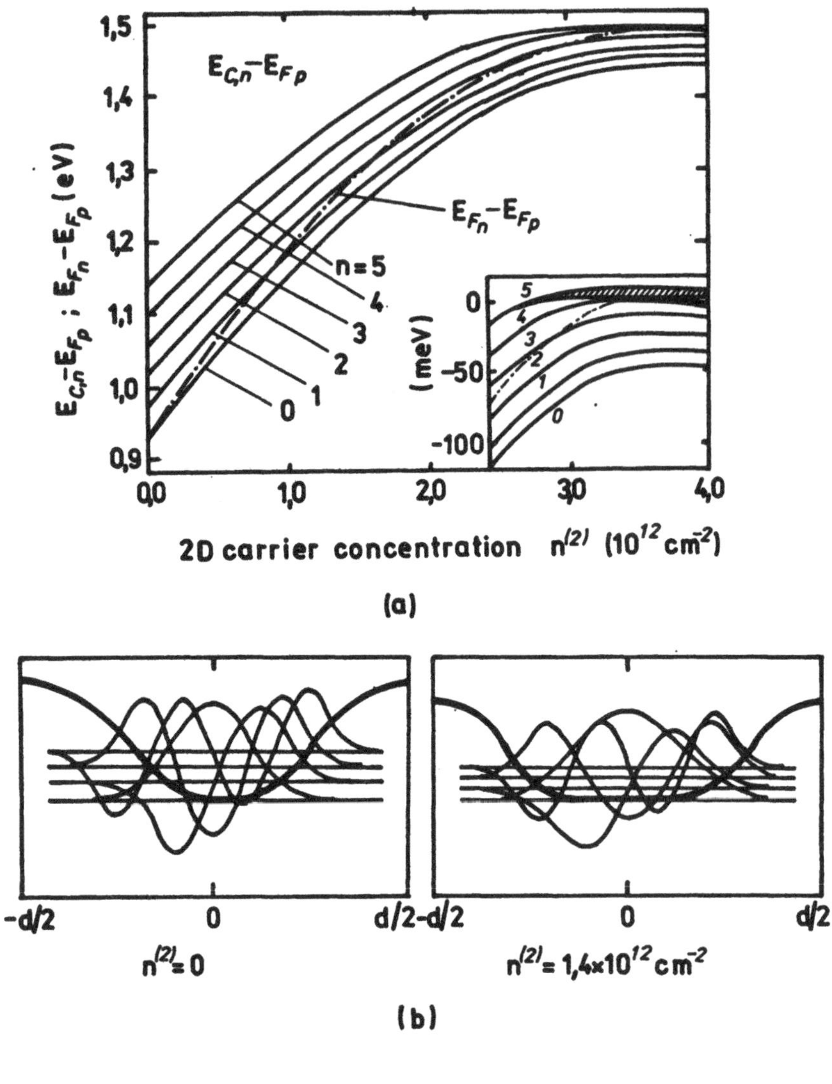

Fig.3.10. Subband energies and the electron quasi-Fermi level as a function of the two-dimensional electron concentration per period (a), and the self-consistent potential energies of the conduction subbands and envelope wave functions for one period (b) of a GaAs n-p doping superlattice (after Ruden et al., Ref. 22).

tration the subband distances decrease monotonously as the shape of the self-consistent potential

$$V_{sc}(z) = V(z) + V_H(z) + V_{xc}(z) \qquad (3.5.25)$$

flattens and broadens due to screaning of the fixed-impurity space charge (see,Fig.3.10b). In order to show the behaviour of the subbands, in particular the finite width of the upper bands at large carrier concentrations, the subband energies are displayed more clearly once more in Fig.3.10a, as an inset on an expanded energy scale and with the maximum of $V_{sc}(z)$, i.e., the classical free-particle threshold, as the zero of energy. The bandwidth becomes appreciable only when the energies nearly reach this treshold. For energies above the treshold, however, the subband gap becomes rather narrow. Thus the subbands change their character from tight-binding to nearly-free electron behaviour within a few meV. This is to be expected because of the wide barrier between neighbouring quantum wells. Another detail which should be noted is the position of E_{F_n} at $n^{(2)} = 4.10^{12}cm^{-2}$ (which corresponds classically to a neutralization of all donors and acceptors by electrons and holes, respectively), which is found to be only slightly above the threshold energy. This means that the electrons almost exactly fill up to the top the self-consistent potential well formed by the impurity charges and the free carriers together.

The corresponding results for a nipi-crystal structure with very thin doping layers of only 4 nm and a thickness of 36 nm for the intrinsic layers are given in Fig. 3.11a. The doping concentrations $N_D = N_A = 5.25\cdot10^{18}cm^{-3}$ have been chosen such that the effective gap has approximately the same value for $n^{(2)} = 0$ as in the former case. From this figure one can see that the subband distances are now larger at small n-values and that they decrease at higher n since the bare potential V(z) has a triangular rather than a parabolic shape. The relation between $E_{F_n} - E_{F_p}$ and $n^{(2)}$ is quite close to a linear dependence because of the wide nearly constant field zones in the intrinsic layers. The subband spacing also decreases with increasing $n^{(2)}$. In contrast to the former case, however, the relative decrease is weaker and affects the higher subbands more than the lower ones: $E_1 - E_0$ decreases only by 1/1.8 between $n^{(2)} = 0$ and $n^{(2)} = N_D d_n$ compared to 1/4.3 in the former case. E_2-E_1, however, decreases now faster than E_1-E_0, na-

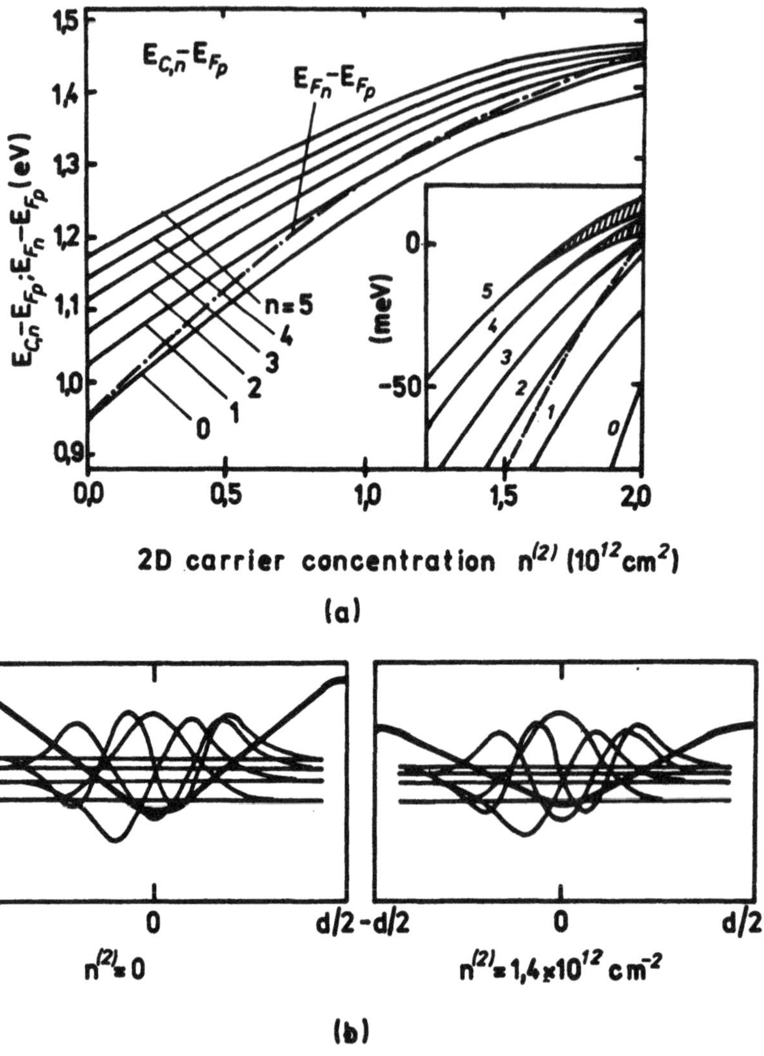

Fig.3.11. Subband energies and the electron quasi-Fermi level as a function of the two-dimensional electron concentration per period (a), and the self-consisted potential, energies of the conduction subbands and envelope wave functions for one period (b) of a GaAs nipi-crystal structure (after Ruden et al.,Ref. 22).

mely by 1/4.2, whereas the corresponding decrease of E_2-E_1 by 1/3.5 was less pronounced for the first example. The two cases exemplify that it is possible to tailor details of the subband structure and of the relation between $E_{F_n}-E_{F_p}$ and $n^{(2)}$ simply by an appropriate choice of the design parameters N_D, N_A, d_n, d_p, and d_i, or by even going one step further and changing from zones of constant doping to continuous doping profiles $N_D(z)$ and $N_A(z)$.

The variation of the bandwidth near the classical free-particle threshold energy is less abrupt for the nipi-case with wide intrinsic layers than for the first example (compare the corresponding insets in Figs. 3.10a and 3.11a), due to the smaller width of the zone with maximum barrier height (see, Figs. 3.10b and 3.11b for a comparison).

A slightly different theoretical formalism for analyzing doping nipi superlattices has been proposed in |93|. This is a modified plane-wave method of the envelope-function approximation type, which is capable of dealing with a spatially variable effective mass. The scheme is particularly easy to implement in a self-consistent manner. The results obtained by using this method have shown, that the effect of the exchange term in the self-consistent potential is small. Although the calculations have been performed in a rather different manner, it appears that the results agree quite well with the experimental and theoretical data available in the literature |3|.

3.6. Literature to chapter 3

|1| Basani,F.: Methods of band calculations applicable to III-V compounds, in: R.K.Willardson, and A.C.Beer (eds.), Semiconductors and semimetals v. 1,21-74, Academic Press, New York 1966

|2| Animalu,A.E.O.: Intermediate quantum theory of crystalline solids, Prentice Hall, Englewood Cliffs 1977

|3| Ploog,K., and G.H.Döhler: Compositional and doping superlattices in III-V semiconductors, Adv.Phys. 32 (1983) 3,285-359

|4| Ivanov,I., and J.Pollmann: Microscopic approach to the quantum size effect in superlattices, Sol.State Commun. 32 (1979) 10,869-872

|5| Ihm,J., P.K.Lam, and M.L.Cohen: Electronic structure of the (001) InAs-GaSb superlattice, Phys.Rev.B 20 (1979) 7,4120-4125

|6| Schulman,J.N., and T.C.McGill: Tight-binding calculation for the AlAs-GaAs (100) interface, J.Vac.Sci.Technol. 15 (1978) 4, 1456-1458

|7| Nucho,R.N., and A.J.Madhukar: Tight-binding study of the electronic structure of the InAs-GaSb (001) superlattice, J. Vac.Sci.Technol. 15 (1978) 4,1530-1534

|8| Schulman,J.N., and T.C.McGill: Electronic properties of the AlAs-GaAs (001) interface and superlattice, Phys.Rev. B 19 (1979) 12,6341-6349

|9| Pollmann,J.: On the electronic structure of semiconductor surfaces, interfaces, and defects at surfaces or interfaces, in: Festkörperprobleme (Adv.Solid State Phys.) v.XX, 117-175, J.Treusch (ed.) , Vieweg, Braunschweig 1980

|10| Pollmann,J., and A.Mazur: Theory of semiconductor heterojunctions, Thin Solid Films 104 (1983) 1,257-276

|11| Sai-Halasz,G.A., L.Esaki, and W.A.Harrison: InAs-GaSb superlattice energy structure and its semiconductor-semimetal transition, Phys.Rev. B 18 (1978) 6,2812-2818

|12| Schulman,J.M., and Y.C.Chang: New method for calculating electronic properties of superlattices using complex band structures, Phys.Rev.B 24 (1981) 8,4445-4448

|13| Chang,Y.C., and J.N.Schulman: Theory of heterostructures- a reduced Hamiltonian method with evanescent states and transfer matrices, J.Vac.Sci.Technol. 21 (1982) 2,540-543

|14| Schulman,J.N., and Y.C.Chang: Reduced Hamiltonian method for solving the tight-binding model of interfaces, Phys.Rev. B 27 (1983) 4,2346-2354

|15| Schulman,J.N., and Y.C.Chang: Band mixing in semiconductor superlattices, Phys.Rev. B 31 (1985) 4,2056-2068

|16| White,S.R., and L.J.Sham: Electronic properties of flat-band semiconductor heterostructures, Phys.Rev.Lett. 27 (1981) 8, 879-882

|17| Bastard,G.: Superlattice band structure in the envelope-function approximation, Phys.Rev. B 24 (1981) 10,5693-5697

|18| Bastard,G.: Theoretical investigations of superlattice band structure in the envelope-function approximation, Phys.Rev.B 25 (1982) 12,7584-7597

|19| Altarelli,M.:Electronic structure of semiconductor superlattices, Physica B/C 117-118 (1983) 2,747-749

|20| Altarelli,M.: Electronic structure and semiconductor-semimetal transition in InAs-GaSb superlattices, Phys.Rev. B 28 (1983) 2,842-845

|21| Voisin,P., G.Bastard, and M.Voss: Optical selection rules in superlattices in the envelope-function approximation, Phys. Rev. B 29 (1984) 2,935-941

|22| Ruden,P., and G.H.Döhler: Electronic structure of semiconductors with doping superlattices, Phys.Rev. B 27 (1983) 6, 3538-3546

|23| Hohenberg,P., and W.Kohn: Inhomogeneous electron gas, Phys. Rev. 136 (1964) 3, B 864- B 871

|24| Kohn,W., and L.Sham: Self-consistent equations including exchange and correlation effects, Phys.Rev. 140 (1965) 4, A 1133- A 1138

|25| Döhler,G.H., and P.Ruden: Properties of nipi doping superlattices in III-V and IV-VI semiconductors, Surface Sci. 142 (1984) 1/3,474-485

|26| Abeles,B., T.Tiedje, H.C.Stasiewski, H.W.Deckman, P.D.Persans, K.S.Liang, and C.B.Roxlo: Amorphous semiconductor superlattices, Superlattices and Microstructures 1 (1985) 2,115-118

|27| Abeles,B., and T.Tiedje: Amorphous semiconductor superlattices, Phys.Rev.Lett. 51 (1983) 21, 2003-2006

|28| Tiedje,T., B.Abeles,P.D.Persans, B.G.Brooks, and G.D.Cody: Bandgap and resistivity of amorphous semiconductor superlattices, J.Non-Crys.Solids 66 (1984) 1/2,345-350

|29| Chang,Y.C., and J.N.Schulman: Interband optical transitions in GaAs-Al_xGa_{1-x}As and InAs-GaSb superlattices, Phys.Rev.B (1985) 4,2069-2079

|30| Bassani,F., and G.P.Parravicini: Electronic states and optical transitions in solids, chap.7, Pergamon Press, New York 1975

|31| Voisin,P.: Optical studies of semiconductor superlattices, Surface Sci. 142 (1984) 2,460-473

|32| Dingle,R.: Confined carrier quantum states in ultrathin semiconductor heterostructures, in: Festkörperprobleme (Adv.Solid State Phys.) v.XV, 21-48, H.J.Queisser (ed.) , Pergamon-Vieweg, Braunschweig 1975

|33| Cardona,M., and F.H.Pollak: Energy band structure of Ge and Si: the $\vec{k}\cdot\vec{p}$ method, Phys.Rev. 142 (1966) 2,530-543

|34| Miller,R.C., D.A.Kleinman, W.A.Nordland, Jr, and A.C.Gossard: Luminescence studies of optically pumped quantum wells in GaAs-Al_xGa_{1-x}As multilayer structures, Phys.Rev. B 22 (1980) 2,863-871

|35| Weisbuch,C., R.C.Miller, R.Dingle, A.C.Gossard, and W.Wiegmann: Intrinsic radiative recombination from quantum states in GaAs-Al_xGA_{1-x}As multi-quantum well structures, Solid State Commun. 37 (1980) 3, 219-222

|36| Gonzalez de la Cruz, G., A.Tselis, and J.J.Quinn: Collective modes in semiconductor superlattices, J.Phys.Chem.Solids 44 (1983) 8,807-812

|37| Das Sarma,S., and J.J.Quinn: Collective excitations in semiconductor superlattices, Phys.Rev. B 25 (1982) 12,7603-7617

|38| Ando,T., A.B.Fowler, and F.Stern: Electronic properties of two-dimensional systems, Rev.Mod.Phys. 54 (1982) 2,437-672

|39| Störmer,H.L.: Novel physics in two dimensions with modulation doped heterostructures, Surface Sci. 142 (1984) 1/3,130-146

|40| Tselis,A.C., and J.J.Quinn: Theory of collective excitations in semiconductor superlattice structures, Phys.Rev. B 29 (1984) 6,3318-3335

|41| Tselis,A.C., G.Gonzalez de la Cruz, and J.J.Quinn: Collective modes of type II semiconductor superlattices, Surface Sci, 142 (1984) 1/3,438-441

|42| Sooryakumar,R., A.Pinczuk, A.C.Gossard, and W.Wiegmann: Dispersion of collective intersubband excitations in semiconductor, superlattices, Phys.Rev. B 31 (1985) 4, 2578-2580

|43| Ruden,P., and G.H.Döhler: Electronic excitations in semiconductors with doping superlattices, Phys.Rev. 27 (1983) 6, 3547-3553

|44| Ehrenreich,H., and M.H.Cohen: Self-consistent field approach to the many-electron problem, Phys.Rev. 115 (1959) 4, 786-790

|45| Tselis,A., and J.J.Quinn: Collective modes of semiconducting space charge layers, Surface Sci. 113 (1982) 1/3, 362-366

|46| Dingle,R., W.Wiegmann, and C.H.Henry: Quantum states of confined carriers in very thin $Al_xGa_{1-x}As$-GaAs-$Al_xGa_{1-x}As$ heterostructures, Phys.Rev.Lett. 33 (1974) 14,827-830

|47| Schulman,J.N.: Effects of compositional grading on GaAs-$Al_xGa_{1-x}As$ interface and quantum well electronic structure, J.Vac.Sci.Technol. B1 (1983) 3,644-647

|48| Messiah,A.: Quantum mechanics, v.1,chap.3, North-Holland P.C., Amsterdam 1967

|49| Schiff,L.I.: Quantum mechanics, 2nd ed., §9, McGraw-Hill B.C. Inc., New York 1955

|50| Price,P.J., and F.Stern: Carrier confinement effects, Surface Sci. 132 (1983) 1/3,577-593

|51| Stern,F., and S.Das Sarma: Electron energy levels in GaAs-$Al_xGa_{1-x}As$ heterojunctions, Phys.Rev.B 30 (1984) 2,840-848

|52| Miller,R.C., A.C.Gossard, D.A.Kleinman, and O.Munteanu: Parabolic quantum wells with the GaAs-$Al_xGa_{1-x}As$ system, Phys. Rev.B 29 (1984) 6,3740-3743

|53| Döhler,G.H.: Ultrathin doping layers as a model for 2D systems, Surface Sci. 73 (1978) 1,97-105

|54| Miller,R.C., D.A.Kleinman, and A.C.Gossard: Energy-gap discontinuities and effective masses for GaAs-$Al_xGa_{1-x}As$ quantum wells, Phys.Rev. B 29 (1984) 12,7085-7087

|55| Stern,F., and J.N.Schulman: Calculated effects of interface grading in GaAs-$Al_xGa_{1-x}As$ quantum wells, Superlattices and Microstructures, 1 (1985) 4,303-305

|56| BenDaniel,D.J., and C.B.Duke: Space-charge effects on electron tunneling, Phys.Rev. 152 (1966) 2,683-692

|57| Bastard,G.: Electronic energy levels in semiconductor quantum wells and superlattices, Superlattices and Microstructures, 1 (1985) 5,265-273

|58| Miller,R.C., D.A.Kleinman,W.T.Tsang, and A.C.Gossard: Observation of the excited level of excitons in GaAs quantum wells, Phys.Rev. B 24 (1981) 2,1134-1136

|59| Mendez,E.E., G.Bastard, L.L.Chang, L.Esaki, H.Morkoc, and R. Fisher: Effect of an electric field on the luminescence of GaAs quantum wells, Phys.Rev. B 26 (1982) 12,7101-7104

|60| Wood,T.H., C.A.Burrus, D.A.B.Miller, D.S.Chemla, T.C.Damen, A.C.Gossard, and W.Wiegmann: High-speed optical modulation with GaAs/AlGaAs quantum wells in a p-i-n diode structure, Appl.Phys.Lett. 44 (1984) 1, 16-18

|61| Alibert,C., S.Gaillard,J.A.Brum, G.Bastard, P.Frijlink, and M.Erman: Measurements of electric-field-induced energy level shifts in GaAs single quantum wells using electroreflectance, Solid State Commun. 53 (1985) 5,457-460

|62| Bastard,G., E.E.Mendez, L.L.Chang, and L.Esaki: Variational calculations on a quantum well in an electric field, Phys. Rev.B 28 (1983) 6,3241-3245

|63| Brum,J.A., and G.Bastard: Electric-field-induced dissociation of excitons in semiconductor quantum wells, Phys.Rev. B 31 (1985) 6,3893-3898

|64| Voisin,P., Y.Guldner, J.P.Vieren, M.Voos, C.Benoit a la Guillaume, N.J.Kawai, L.L.Chang, and L.Esaki: Optical studies of InAs-GaSb superlattices, J.Phys.Soc.Jap. 49 (1980) Suppl. A,1005-1008

|65| Bastard,G.: Hydrogenic impurity states in a quantum well: A simple model, Phys.Rev. B 24 (1981) 8,4714-4722

|66| Shanabrook,B.V., and J.Comas: Photoluminescence from "spike doped" hydrogenic donors in $Al_{0.3}Ga_{0.7}As$-GaAs quantum wells, Surface Sci. 142 (1984) 1/3,504-508

|67| Jarosik,N.C., B.D.McCombe, B.V.Shanabrook, J.Comas, J.Ralston, and G.Wicks: Binding of shallow donor impurities in quantum well structures, Phys.Rev.Lett. 54 (1985) 12,1283-1286

|68| Priester,C., G.Allan, and M.Lannoo: Validity of the effective mass approximation for shallow impurity states in narrow superlattices, Phys.Rev.B 28 (1983) 12,7194-7200

|69| Priester,C., G.Allan, and M.Lannoo: Resonant impurity states in quantum well structures, Phys.Rev. B 29 (1984) 6,3408-3411

|70| Masselink,W.T., Y.C.Chang, and H.Morkoc: Binding energies of acceptors in GaAs-$Al_xGa_{1-x}As$ quantum wells, Phys.Rev.B 29 (1983) 12,7373-7376

|71| Masselink,W.T., Y.C.Chang, and H.Morkoc: Binding energies of acceptors in GaAs-$Al_xGa_{1-x}As$ quantum wells, J.Vac.Sci.Technol. B2 (1984) 3,376-382

|72| Smith,R.A.: Wave mechanics of crystalline solids, chap.4, J.Wiley, New York 1961

|73| Kittel,C.: Introduction to solid state physics, 5th ed., 280-283, J.Wiley, New York 1976

|74| Esaki,L., and R.Tsu: Superlattice and negative differential conductivity in semiconductors, IBM J.Res.Develop. 14 (1970) 1,61-65

|75| Esaki,L., L.L.Chang, W.E.Howard, and V.L.Rideout: Transport properties of a GaAs-AlGaAs superlattice, in: Proceedings of the XI Intern.Conf.Phys.Semicond.,Warsaw, Poland 1972, v.1, 431-437, PWN-Polish Scientific Publishers, Warsaw, and Elsevier, Amsterdam 1972

|76| Mukherji,D., and B.R.Nag: Miniband parameters of semiconductor superlattices, Solid State Electronics 18 (1975) 12, 1107-1109

|77| Mukherji,D., and B.R.Nag: Band structure of semiconductor superlattices, Phys.Rev.B 12 (1975) 10,4338-4345

|78| Ando,T. and S.Mori: Effective-mass theory of semiconductor heterojunctions and superlattices, Surface Sci. 113 (1982) 1,124-130

|79| Milanovic,V., and D.Tjapkin: Energy band calculation and zero energy gap conditions for semiconductor superlattices, Phys.Stat.Sol. (b) 110 (1982) 2,687-695

|80| Chadi,D.J., and M.L.Cohen: Tight-binding calculations of the valence bands of diamond and zincblende crystals, Phys.Stat. Sol. (b) 68 (1975) 1,405-419

|81| Mori,S., and T.Ando: Electronic properties of a heavily doped n-type GaAs-$Al_xGa_{1-x}As$ superlattice, Surface Sci. 98 (1980) 1,101-107

|82| Matthews,J.W., and A.E.Blakeslee: Almost perfect epitaxial multilayers, J.Vac.Sci.Technol. 14 (1977) 4,989-991

|83| van der Merve,J.H.: Crystal interfaces. Part II-Finite overgrowths, J.Appl.Phys. 34 (1963) 1,123-127

|84| Osbourn,G.C.: Strained-layer superlattices from lattice mismatched materials, J.Appl.Phys. 53 (1982) 3,1586-1589

|85| Chadi,D.J.: Spin-orbit splitting in crystalline and compositionally disordered semiconductors, Phys.Rev.B 16 (1977) 2, 790-796

|86| Chelikowsky,J.R., and M.L.Cohen: Nonlocal pseudopotential calculations for the electronic structure of eleven diamond and zincblende semiconductors, Phys.Rev.B 14 (1976) 2,556--582

|87| Camphausen,Don L., G.A.Neville Connel, and W.Paul: Calculation of energy-band pressure coefficients from the dielectric theory of the chemical bond, Phys.Rev.Lett. 26 (1971) 4, 184-188

|88| Osbourn,G.C.: $In_xGa_{1-x}As$-$In_yGa_{1-y}As$ strained-layer superlattices-a proposal for useful, new electronic materials, Phys. Rev.B 27 (1983) 8,5126-5128

|89| Dodson,B.W.: Stability of registry in strained-layer superlattice interfaces, Phys.Rev. B 30 (1984) 6,3545-3546

|90| Osbourn,G.C.: Electron and hole effective masses for two-dimensional transport in strained-layer superlattices, Superlattices and Microstructures 1 (1985) 3,223-226

|91| Kane,E.O.: Band structure of indium antimonide, J.Phys.Chem. Solids 1 (1957) 4,249-261

|92| Chang,L.L., and L.Esaki: Electronic properties of InAs-GaSb superlattices, Surface Sci. 98 (1980) 1,70-89

|93| Brand,S., and R.A.Abram: Self-consistent calculations of electron and hole subband energies for an n-p superlattice in GaAs, J.Phys.C: Solid State Phys. 16 (1983) 29,6111-6120

Chapter 4. EXPERIMENTAL RESULTS ON SEMICONDUCTOR SUPERLATTICES

Experimental investigations on superlattices were started in 1972 when the first compositional GaAs-$Al_xGa_{1-x}As$ superlattices grown by MBE become available |1|,|2|. Since that time, many of the fundamental predictions made by different researchers |3-8| could be verified. In this chapter some selected experimental results will be described, which are related to the structural parameters, to optical transitions and to electrical transport in different types of superlattices. The research field related to superlattices is growing very fast, therefore, the data presented in this chapter should not be treated as an exhaustive survey of the relevant literature, but rather as a general view on the subject.

4.1. Structural parameters of superlattice structures

Smooth growth is obviously a prerequisite for the exploration of the characteristic quantum effects in superlattice structures. It has been recognized that the MBE, as well as the MO CVD processes produce extremely smooth growths of III-V and II-VI compound semiconductors under optimum growth conditions. These results presumably arise from the fact that the step propagation process is dominant during the entire crystal growth if contaminations are kept minimal on the surface.

Among various methods employed for structural examination, two techniques, Auger depth profiling |9| and X-ray small angle interference measurements |10|,|11|, have provided most important information with respect to the compositional variations and the interfacial smoothness.

An Auger spectrum of, e.g., a clean $Al_xGa_{1-x}As$ surface indicates the Ga,As, and Al lines over the 1000-1400 eV range. The intensity ratio of the 1390 eV Al signal to the 1228 eV As signal may be calibrated as a function of the Al concentration, predetermined by Raman spectroscopy |12|. With this calibration, the composition profile of the multilayer structure can be obtained by the simultaneous use of ion sputter-etching of the specimen surface and Auger electron spectroscopy |13-15|. Prior to the profiling experiment, the ion pump of the ultrahigh vacuum chamber, where the investigated specimen is placed, is turned off and argon gas blend into the chamber to a pressure of $(1.5\text{-}2.5)\cdot10^{-5}$ Torr. With an argon ion energy of 1.5 keV, sputter-etching rates for GaAs of 0.3-1 nm/min have been measured, based upon the known thickness of the layers. The three lines (Ga,As and Al) are sequentially scanned during the sputtering operation. Fig. 4.1a shows the result of the compositional profile of a periodic structure consisting

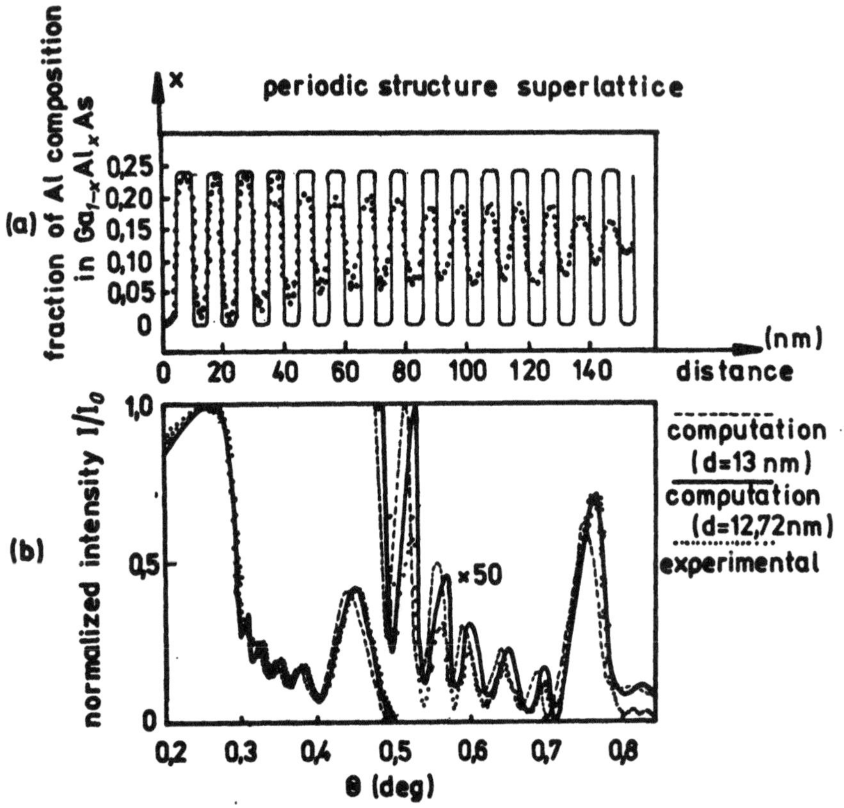

Fig.4.1.a) Composition profile of a GaAs-AlGaAs superlattice structure measured by a combination of ion sputter-etching and Auger electron spectroscopy, and b) X-ray interference pattern for a GaAs-AlAs superlattice with six periods of 12.72 nm. In comparison with the experimental points shown in (b) computer results with different period thicknesses are plotted to show the agreement (full curve) and the discrepancy (broken curve)(after Esaki et al.,Ref.12).

of alternating 5 nm layers of GaAs and $Al_{0.25}Ga_{0.75}As$. The steady decrease in the peak-to valley ratio with increasing depth of removed material is attributed to spatially varying removal rates on the scale of the area of the focused primary electron beam (typically 100 μm in diameter). The possibility that the smearing of the composition is due to diffusion effects has been eliminated. Since the sampling depth or the escape depth for the 1390 eV Auger electrons is in the 1-2 nm range, the actual profile is believed to be the one which is shown by the solid line in this figure.

Small and large angle X-ray scattering measurements give another possibility for the evaluation of the superlattice structure. In the former, X-ray interference patterns due to multilayer structures can be observed, whereas in the latter, in addition to the Bragg reflection peaks due to the host crystals, a number of satellite peaks arising from the superlattice periodicity are usually resolved. Small angle X-ray interference measurements performed using a computer-controlled diffractometer with Cu $K\alpha_1$ radiation (λ=0.154 nm) on specimens which have a GaAs-AlAs periodic structure with only six periods, formed near the surface of GaAs substrates, gave a typical interference pattern shown in Fig. 4.1b. The radiation was monochromated by double reflection at a grooved Ge crystal and collimated by the use of a series of slits. The beam size at the sample had a length of 2 mm or 5 mm depending on the slit used, and a width of 0.05 mm/sinθ, where θ is the incident angle with respect to the sample surface. Figure 4.1b shows the interference pattern, taken with a 5 mm slit, for the central portion of a sample with a period consisting of an intended layer thickness of 7,5 nm GaAs and 5 nm AlAs as predicted from the growth rates. The critical angle for total reflection at which the intensity drops drastically occurs near 0.3°. The gradual decrease in intensity below this angle results from incomplete collection of the totally reflected beam because of the finite sample size. Beyond the critical angle, strong interference peaks are observed: two principal peaks at approximately 0.45° and 0.75° and a number of secondary peaks. Similarly well-defined patterns can be observed in structures with different periods. The pattern arises from the difference in refractive indices for X-rays between GaAs and AlAs in the periodic structure, analogous to that in optical diffraction involving a series of regularly spaced slits.

The experimental data shown in Fig. 4.1b have been compared quantitatively with theoretical calculations using the recurrence formula derived for a multilayer structure based on the continuity of tangential components of the electric vectors at the interfaces. The results calculated to fit the data points are shown as a full curve for a period thickness of 12.72 nm with a GaAs thickness of 7.6 nm and an AlAs thickness of 5.12 nm, which are close to those intended. The two interference patterns of the data points and the calculation agree very well not only in angular positions but in intensities and line-widths. It is important to note that a distinctly different calculated pattern is obtained by changing the period by 0.28 nm to d = 13 nm, while keeping other parameters fixed. This is shown as a broken curve in the figure for the same GaAs layer of 7.6 nm but an AlAs layer of 5.4 nm. In fact, even a change of one atomic spacing [0.141 nm for the (100) surface] makes a distinguishable difference. It is reasonable to believe that this result gives direct evidence of the achievement of an interfacial smoothness and a periodic coherency on the scale of atomic spacings.

X-ray diffraction measurements may also be used to investigate the interdiffusion |16| and the impurity diffusion |17| in superlattices. X-ray diffraction from a material in which the lattice parameter and/or the scattering power is subject to a one-dimensional modulation is characterized by satellites around each Bragg peak of the average lattice. The intensities of the satellite peaks are proportional to the Fourier components of the direct space composition modulation. Therefore, the interdiffusion of, e.g., GaAs and AlAs in a GaAs/AlAs superlattice can be studied by measuring the satellite intensities as a function of annealing time at a constant temperature. In addition, measurements of the widths of the satellites in reciprocal space yield information about the overall quality of the films such as the coherence of the multilayer stacking sequence both parallel and perpendicular to the modulation wave vector.

4.1.1. X-ray rocking curve analysis of superlattices

A slight modification of the X-ray interference method of structural examination of superlattice structures described above, leads to the double-crystal X-ray rocking curve method |10|,|18-23|.

To record rocking curves the X-ray beam is first reflected from the (hkl) planes of a well polished block of substrate material and then from the $(\bar{h}\bar{k}\bar{l})$ planes of the sample. Slits are used to confine the area of sample irradiated. The width of the rocking curve, though not the area under it, is a function of the relative tilt of the Bragg planes about an axis in the incidence plane and perpendicular to the reflecting plane normal. Once the angular position for diffraction from both crystals has been found, the second crystal is tilted to bring the reflecting plane normal into the plane of incidence of the first crystal. This procedure is iterative and involves the sequential measurement of rocking curve peak height as the second crystal is tilted. At maximum peak height, the Bragg planes are parallel and the full width at half maximum (FWHM) is minimised. It is then independent of the source and camera geometry.

The rocking curve recorded, for example, when the second crystal has a single epitaxial layer usually consists of two distinct peaks. The lattice constant difference between layer and substrate is calculated from the peak separation and the ratio of the integrated intensities of the two peaks gives a measure of layer thickness. In addition, the peak shapes give information about the crystalline perfection of the layer.

This is one of the most sensitive methods available to measure compositional and thickness variations. Lattice constant differences can be measured to within 10 parts per million and 10% variations in layer thickness should be readily detectable for good quality layers. For ternary compounds layers, the double crystal diffractometry is able to characterise the layer completely since the composition can be calculated directly from the lattice constant. The composition is determined using Vegard's law and assuming values for the X-ray wavelengths and the lattice parameter (the lattice constant) of the constituent binary compounds. Although there is no evidence to suggest that there is any nonlinearity in the lattice constant variation with composition, any such deviations do not render rocking curve results less useful. This is because the important factor governing the growth of epitaxial layers of high crystalline quality is the attainment of a close lattice match between layer and substrate. This is checked directly using X-ray measurements and the absolute composition is of secondary importance. The same is true of quaternary compounds layers.

Layer thickness variations can be qualitatively deduced from peak area ratios. For layers of good crystalline perfection quantitative estimates are possible by comparing experimental and theoretical rocking curves. The precision achievable is comparable with other methods involving the microscopic examination of cleaved sections.

The X-ray rocking curve method enjoys a number of advantages over ion channeling, Rutherford backscattering, electron diffraction, and Auger electron spectroscopy, all which have been used to measure properties of superlattices. The complexity and cost of the apparatus are an order of magnitude lower. The measurement of rocking curves is simple, rapid, and reproducible. It is also nondestructive. Moreover for measurement of strain profiles, the rocking curve is unmatched by either channeling or electron diffraction. Following the considerations of Ref. 20 we will discuss now this technique in application to superlattices. The technique measures depth profiles of structure factor, and profiles of perpendicular and parallel strains relative to the underlying substrate. In addition to providing a detailed picture of the state of stress, the profiles are a direct measure of the composition modulation. The thickness of the period of modulation and the average strain are determined with a precision of about 1%. The detailed structure of the period is determined to about 5%.

When considering the reflecting power of a superlattice to inciding X-rays, it is convenient to define first the X-ray strains of an epitaxial film with respect to the substrate, since these are determined directly from the rocking curve. Denoting the film and the substrate by f and s, respectively, for an arbitrary set of planes, there is, in general, a difference $\Delta d = d_f - d_s$ in interplanar d spacing. The difference depends on the particular deformation as well as on the planes. In principle, the strain ε is a tensor with unequel normal and shear components. For cubic crystals of arbitrary orientation, such as <311>, it is necessary to consider shear strains. But for layers grown along the <100>, <110>, or <111> directions, the principal strains are perpendicular and parallel to the layer. In these directions, the X-ray strains $\varepsilon_\perp$ and $\varepsilon_\parallel$ are the fractional differences in interatomic spacing between the film and the substrate. The strains are related to the difference in d spacing through

$$\Delta d/d_s = \varepsilon_\perp \cos^2\psi + \varepsilon_\parallel \sin^2\psi \qquad (4.1.1)$$

where ψ is the angle between the planes and the surface. For perfectly coherent epitaxy, $\varepsilon_\parallel = 0$ regardless of the mismatch between the free film and the free substrate. Even under partial relaxation ($\varepsilon_\parallel \neq 0$), this condition imposes stresses of opposite signs in the layer and substrate, and in the simplest case produces tetragonal distortions and bending. Second-order variations of $\varepsilon_\parallel$ with direction in the plane of the layer occur for nonorthotropic orientations. Since the thickness of the substrate is usually two orders of magnitude greater than the thickness of the layer, the strains in the substrate are usually two orders of magnitude smaller than those in the film. To a good approximation, the substrate is unstrained and one may substitute the free substrate interplanar spacing d_s^o for d_s in Eq. (4.1.1). If the free-lattice parameter of the film is known, it is a simple matter to convert the X-ray strain, defined relative to the substrate, to the strain of elasticity theory, defined relative to the free film.

For diffraction calculations, uniform epitaxial layer is described by its thickness t, structure factor F, perpendicular and parallel strains $\varepsilon_\perp$ and $\varepsilon_\parallel$, and normal absorption coefficient μ. With respect to the inward normal to the surface, the direction cosines of the incident and diffracted waves are γ_o and γ_H, respectively. The angle between the diffracting planes and the surface is ψ. Associated with the epitaxial layer and the particular reflection are the quantities A and Y defined as

$$A = (r_e\lambda|F|t)/(V\sqrt{|\gamma_o\gamma_H|}) \qquad (4.1.2)$$

$$Y = -\sqrt{\gamma_o/|\gamma_H|}\,(\pi V \sin 2\theta_B)/(r_e\lambda^2|F|)(\Delta\omega) \qquad (4.1.3)$$

where r_e is the classical electron radius, λ is the X-ray wavelength, V is the volume of the unit cell, and θ_B is the Bragg angle of the substrate. The differential angle $\Delta\omega$ is equal to

$$\Delta\omega = \theta-\theta_B + (\varepsilon_\perp \cos^2\psi+\varepsilon_\parallel \sin^2\psi)\cdot\tan\theta_B \pm (\varepsilon_\perp - \varepsilon_\parallel)\sin\psi\cos\psi \qquad (4.1.4)$$

where θ is the grazing angle of incidence with respect to the diffracting planes. The + or - sign is chosen according to whether the angle of incidence with respect to the surface is $\theta_B-\psi$ or $\theta_B+\psi$, respectively. In Eq. (4.1.4), the first term involving strains represents a change in d spacing while the second is due to the rotation of the planes.

An arbitrary depth profile of strains and structure factor can, for computational convenience and with sufficient accuracy, be represented by a discrete structure of N laminae. The normalized amplitude diffracted by such a structure is

$$E_N = i\sqrt{\gamma_o/|\gamma_H|}\sum_{j=1}^{N} a_j \exp[-i(A_y A_y+\phi_j)]\sin(A_j Y_j)/Y_j \qquad (4.1.5)$$

where

$$a_j = \exp[-\mu(\gamma_o+|\gamma_H|)/(2|\gamma_o\gamma_H|)\sum_{i=j+1}^{N} t_i]$$

$$a_N = 1$$

$$\phi_j = 2\sum_{i=1}^{j-1} A_i Y_i \ , \quad \phi_1 = 0$$

and each lamina j has its own A_j and Y_j. A superlattice is a special case of the arbitrary laminar structure. In its simplest form, the superlattice period consists of two layers, labeled a and b, each with its own thickness, strains, structure factor, and the corresponding A_a, Y_a and A_b, Y_b. For M superlattice periods and neglecting normal absorption, the diffracted amplitude [Eq. (4.1.5)] becomes

$$E_M = i\sqrt{\gamma_o/|\gamma_H|}\exp(-i\beta)F_s \sin[M(A_aY_a+A_bY_b)]\,[\sin(A_aY_a+A_bY_b)]^{-1} \qquad (4.1.6)$$

where

$$\beta = (M-1)(A_aY_a+A_bY_b) + A_aY_a$$

and

$$F_s = \sin(A_aY_a)/Y_a+\exp[-i(A_aY_a+A_bY_b)]\sin(A_bY_b)/Y_b \qquad (4.1.7)$$

The quantity F_s can be regarded as the structure factor of one superlattice period. The sinusoidal term in the numerator of Eq. (4.1.6) produces zeros with a period $\Delta\theta_M$ given by

$$M\Delta(A_aY_a+A_bY_b) = \pi$$

or

$$\Delta\theta_M = (\lambda|\gamma_H|)/[M(t_a+t_b)\sin 2\theta_B] \qquad (4.1.8)$$

from which the total thickness $M(t_a+t_b)$ of the superlattice can be determined. However, for typical samples (≈1 μm thick) lateral nonuniformities frequently result in a convolution of the rapid oscillations. The more slowly varying sinusoidal term in the denominator of Eq. (4.1.6) produces observable peaks at

$$A_aY_a + A_bY_b = n\pi \qquad (4.1.9)$$

hence, the peaks are labeled 0,±1,±2, etc. The spacing $\Delta\theta_p$ between the peaks yields the superlattice periodicity p, equal to

$$p = t_a+t_b = (\lambda|\gamma_H|)/(\Delta\theta_p \sin 2\theta_B) \qquad (4.1.10)$$

The superlattice zeroth-order peak is located at an angle $\Delta\theta_o$ from the substrate peak. From Eq. (4.1.9) one obtains

$$-\Delta\theta_o = k_1\langle\varepsilon_\perp\rangle + k_2\langle\varepsilon_\parallel\rangle \ ,$$

$$k_1 = \cos^2\psi \tan\theta_B \pm \sin\psi \cos\psi \qquad (4.1.11)$$

$$k_2 = \sin^2\psi \tan\theta_B \mp \sin\psi \cos\psi \ ,$$

where <> denotes depth averaging (over the superlattice period) and the sign is chosen as described above. The amplitude of the n-th order peak is proportional to the superlattice structure factor [Eq.(4.1.7)] evaluated at angles determined by Eq. (4.1.9). Using n as a subscript one can write

$$F_{sn} = \sin(A_aY_{an})[Y_{an}^{-1} - Y_{bn}^{-1}] = \sin(A_aY_{an})/(A_aY_{an})\{A_o +$$

$$+ A_b/[1-n\pi/(A_aY_{an})]\} \qquad (4.1.12)$$

Since Y_{an} is related to Y_{a0} through Eqs. (4.1.10) and (4.1.3), Eq. (4.1.12) depends on A_a,A_b, and A_{a0} only. For n=0, this reduces to

$$F_{s0} = \sin(A_aY_{a0})/(A_aY_{a0})[A_a+A_b] \simeq (A_a+A_b) \ , \ A_aY_{a0} \ll 1 \qquad (4.1.13)$$

Thus, the amplitude of the zeroth-order peak measures $F_at_a+F_bt_b$, where F_a and F_b are the crystallographic structure factors of layers a and b, respectively. The approximation in Eq. (4.1.13)is quite good for typical strong reflections and products of strain modulation and period thickness below ~1%·30 nm. In these cases, the zeroth-order peak is more intense than higher-order peaks. For larger strains and thicknesses, the approximation in Eq. (4.1.13) breaks down and the zeroth-order peak may be less intense than higher-order peaks.

From Eq. (4.1.6) it is clear that the basic structure of the rocking curve is determined by the structure of the superlattice period. The number of peaks, their location, and relative amplitudes are independent of the number of periods, provided this number is greater than 1. For M periods and negligible absorption, the overall intensity of the rocking curve goes as M^2. A perfect superlattice is described by six parameters: t_a, t_b, ε_a, ε_b, F_b/F_a and M. To determine these parameters, one can use Eqs. (4.1.8), (4.1.10),(4.1.11), and (4.1.13), and Eq. (4.1.12) evaluated for the ±1 peaks. The locations and amplitudes of peaks corresponding to $|n|>1$ are completely determined by the six parameters. To separate the components of ε_a and ε_b, one needs at least two rocking curve measurements. The availability of a large number of intense reflections with varying degrees of asymmetry enables the verification of the internal consistency of the depth profiles of strain and the structure factor. In fact, the present method can be extended to arbitrary deformations, including shear strains. For each additional strain component, an additional rocking curve measurement is needed.

The solution of the six simultaneous equations is straightforward provided the condition in Eq. (4.1.13) is satisfied. Since this is not always the case, it may be necessary to consider higher-order peaks. Alternatively, since the condition of Eq. (4.1.13)depends on the Bragg angle and on ψ as well as on the strain and thickness, it is almost always possible to use a reflection where the zeroth-order peak is the most intense. In practice, the number of periods and the approximate composition modulation are known from the growth conditions. One can calculate a priori F_a and F_b from the nominal composition. This reduces the number of unknowns to four without changing the number of equations (six). Thus, for a perfect or nearly perfect superlattice, we are able to determine its structure by following a specified algorithm and are not dependent on computer fitting.

If the structure of the superlattice period has more detail than the bilayer model assumed above, it is nevertheless clear that the form of Eq. (4.1.6) remains valid. By straightforward extension, Eqs. (4.1.6) and (4.1.7) can describe diffraction in superlattices with an arbitrary number of sublayers in one period. For each additional sublayer, the intensity of an additional high-order peak must be considered. However, Eqs. (4.1.10) and (4.1.11), which determine the periodicity and the average strain, remain applicable regardless of the structure of the period.

The presented theory has been applied first to two different type I superlattices, namely, to the 15 periods, 40 nm per period ($t_a=t_b=p/2$) $Al_{0.9}Ga_{0.1}As/GaAs$ superlattice grown on a (100) GaAs substrate, and to the strained-layer, 10 periods, 60 nm per period, ($t_a=t_b=p/2$) AlSb/GaSb superlattice grown on a (100) GaSb substrate. These two kinds of superlattices are examples of extreme cases of stain variation likely to be found in practice. For the AlGaAs/GaAs superlattice, the maximum strain is below 0.3%, while for the AlSb/GaSb superlattice, the strain modulation is greater than 1%. In both cases, the nominal thickness of the period is large, so that for the AlSb/GaSb superlattice the condition in Eq. (4.1.13) above is not satisfield, i.e., the zeroth--order peak is less intense than higher-order peaks. For the AlGaAs/GaAs superlattice, the zeroth-order peak is the most intense.

Figure 4.2a shows measured (dashed line) and calculated (solid line) Fe K_{α_1} (400) rocking curves of the AlGaAs/GaAs superlattice. In the experimental curve, in addition to the substrate peak (located at zero), seven superlattice peaks are clearly visible. These are obviously not rapid oscillations of Eq. (4.1.8) but peaks given by Eq. (4.1.9). Their spacing yields an average superlattice period thickness p = (67,6±0.2) nm. The location $\Delta\theta_0$ =-0.0641° of the zeroth-order peak yields, through Eq. (4.1.11) $\langle\varepsilon_\perp\rangle$=0.118%. The (400) rocking curve gives no information on $\varepsilon_\parallel$, but previous work has shown |24| that it is zero for epitaxial AlAs layers up to several microns thick, grown on GaAs.

Assuming that the nominal GaAs layer is indeed GaAs, its perpendicular (and parallel) strain must be zero. The task of determining the strain profile in the superlattice period is thus reduced to determining the thickness and strain of the AlGaAs layer. Using the ratio of the amplitude of the +1 peak to that of the zeroth peak and the appropriate structure factors, Eq. (4.1.12) yields a thickness and strain of 32 nm and 0.249%, res-

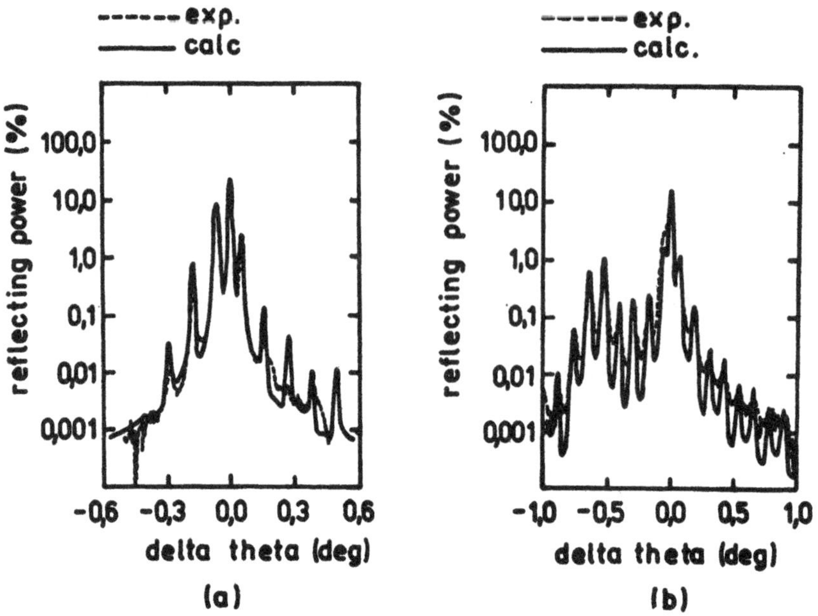

Fig.4.2. Measured (dashed line) and calculated (solid line) Fe Kα (400) rocking curves of AlGaAs/GaAs superlattioe (a), and of the AlSb/GaSb superlattice (b). The calculated curves correspond to the bilayer structure of the superlattice period discussed in the text (after Speriosu et al.,Ref.20)

pectively, for the AlGaAs layer. The same values are obtained using the ratio of the amplitude of the -1 peak to that of the zeroth peak. The agreement between these two determinations shows that the initial assumption of zero strain in the GaAs layer is valid. Had this assumption been false, the +1 and -1 peaks would have yielded different thicknesses and strains for the AlGaAs layer. The calculated rocking curve of Fig. 4.1a, was obtained from this initial strain and structure factor distribution. The calculated curve reproduces very well the locations of the observed peaks as well as the intensities of the three low-order peaks.The slight discrepancy in the intensities of the +1 and -1 peaks is due to the small error in thickness and strain obtained from Eq. (4.1.12) which neglected absorption, whereas the calculated curve in the figure includes normal absorption. For high-order peaks, the calculated curve generally predicts more intensity than is observed. Since random fluctuations in layer thickness were shown to decrease high-order more than low-order peaks, one may suppose that the discrepancy is due to such fluctuations.

Figure 4.2.b shows experimental and calculated Fe $K\alpha_1$ (400) rocking curves for the AlSb/GaSb superlattice. In addition to the substrate peak, the measured curve contains 14 clearly visible superlattice peaks. As mentioned earlier, due to the combination of large strain modulation and large thickness of the period, the zeroth-order peak is less intense than higher-order peaks. This situation presents the practical problem of identifying the zeroth-order peak. The simplest way to resolve the dilemma is to calculate a rocking curve using data based on the nominal composition modulation. The actual thickness of the period (61.0±0.2) nm, is determined directly from the measured curve. Using the appropriate structure factors the nominal thickness ratio of 1:1 in the bilayer model of the period, and assuming zero perpendicular strain in the GaSb layer, one can calculate rocking curves corresponding to various values of perpendicular strain in the AlSb layer. Comparison with the experimental curve immediately shows that the strain in the AlSb layer is around 1.2% while the strain in the GaSb layer is indeed close to zero. Consequently, the zeroth-order peak is at $\Delta\theta_0 \approx -0.29^\circ$ in Fig. 4.2b. This result may be verified by comparing calculated and measured intensities of all peaks. The good agreement shown in this figure was obtained by trial-and-error adjustment in the structure of the period. For this sample, the structure is very nearly a step function. The thickness of the transition region between the two layers is at most 5% of the thickness (30.5 nm) of the layers. The perpendicular strains are (1.25±0.02)% and (-0.03±0.02)% for the AlSb and GaSb layers, respectively. Negative perpendicular strain in the GaSb layer implies positive parallel strain.

In conclusion, it should be noted that the form of the given simple expression relating the structure of the superlattice period to the structure of the rocking curve shows at a glance the existence of equally spaced peaks whose intensities are determined by the structure of the superlattice period. The location and intensity of the zeroth (and frequently larger) order-peak measure the depth-averaged properties of the superlattice. For a perfectly periodic superlattice whose period is a step function, consideration of the three lowest-order peaks provides an analytic determination of the period. For more complicated structures, including transition regions between the layers of the period, one must consider the intensities of higher-order peaks. Small, random fluctuations in layer thickness and strains decrease the intensities of high-order peaks with relatively small changes for low-order peaks. The treatment above includes depth profiles of structure factor and perpendicular and parallel strains.

4.1.2. Ion backscattering and channeling studies on superlattices

The use of Rutherford backscattering and channeling measurements in the analyses of superlattices had begun, when for the first time a set of GaSb/InAs superlattices were studied by backscattering of high-energy (~2 MeV) helium ions |25|,|26|. These studies were followed by investigations performed on GaSb/AlSb |27| and Ge/GaAs |28| superlattices. Recently, detailed compositional and structural analyses of $Al_xGa_{1-x}As$/GaAs superlattices have been carried out by 2MeV-He^+ backscattering with channeling complemented by X-ray rocking curve measurements |29|,|30|. Through the combined use of the two techniques, depth profiles of strain, composition, and crystalline quality could be determined.

Combined with the channeling technique, high-energy ion beams offer an extremely fine and well-characterized probe which is sensitive to lattice strain and dynamics. Since the use of such probe for atomic structure analyses of crystal surfaces and interfaces has been first demonstrated in 1967 |31|, many works have been carried out |32-34|. The principles, techniques, and various applications are summarized in a monograph by Feldman et al. |35| Here we will discuss the problem only briefly using as the example the investigations on the GaSb/AlSb superlattice |27|.
Before starting, let us however, consider first the basic physics of the MeV ion scattering in solids |34|. The geometry shown in Fig.4.3a illustrates a projectile of mass M_1, atomic number Z_1, and energy E_0 scattering from a surface atom of a solid composed of atoms of mass M_2 and atomic number Z_2 and surface impurity M_3, Z_3.

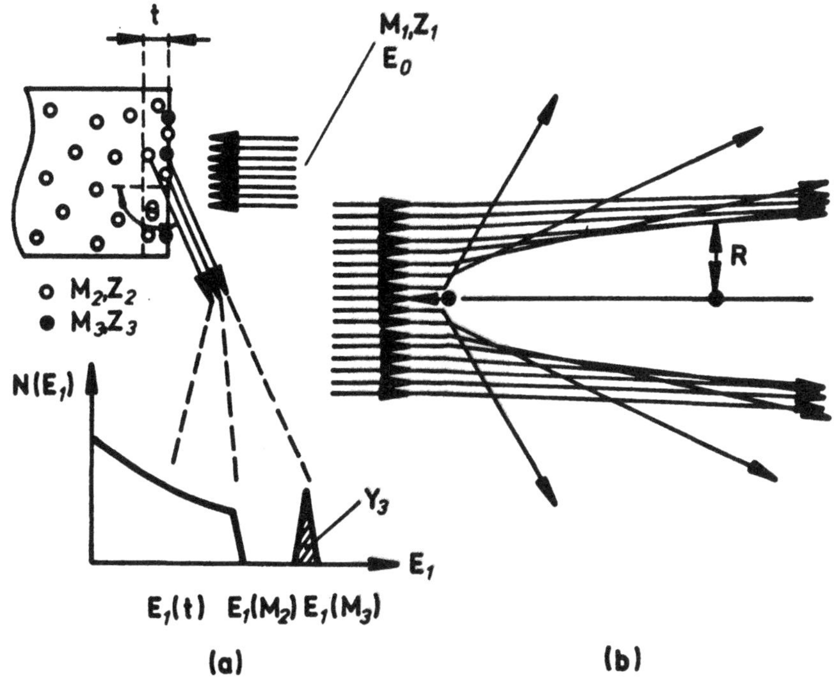

Fig.4.3. a) Geometry for particles of incident energy E_0, mass M_1 and atomic number Z_1 scattering from a surface atom and, an atom at depth t of a solid in which the constituent atoms have mass M_2 and atomic number Z_2, and from a surface impurity atom of mass $M_3 > M_2$ and atomic number Z_3. The energy spectrum of scattered particles with a surface peak of area Y_3 corresponding to the areal density of the species M_3, and a continuum corresponding to scattering from various depths of the substrate.
b) Formation of a shadow cone at the second of a pair of atoms aligned with an incident ion beam. The shadow cone radius at the second atom is denoted R (after Feldman, Ref.34).

The scattering angle is θ and the detector subtends a solid angle $\Delta\Omega$ about the θ direction. The energy E_1 of particles backscattered from a surface atom M_2 is related to the incident energy through the kinematic factor K, where

$$K_{M_2}(\theta) = \{[M_2^2 - M_1^2 \sin^2\theta)^{1/2} + M_1 \cos\theta] / (M_2 + M_1)\}^2 = E_1/E_o \quad (4.1.14)$$

This formula acquires a particularly simple form at θ = 90 and 180°

$$K_{M_2}(90^o) = (M_2-M_1)/(M_2+M_1) \quad (4.1.15)$$

and

$$K_{M_2}(180^o) = (M_2-M_1)^2/(M_2+M_1)^2 \quad (4.1.16)$$

The mass of the scattering atom determines the backscattered energy.

Energetic light ions can penetrate the order of microns into a solid losing their energy primarily through electronic interactions. The rate of energy loss, dE/dx, is a function of the atomic parameters of the projectile and host and a function of incident energy. Extensive tables of dE/dx are available |36|. In the 0.5 - to 2.0-MeV energy region for He^+ projectiles, dE/dx is relatively independent of energy and simple analyses of depth effects are possible. More thorough analyses are not conceptually difficult but may require numerical techniques. The approximate constancy of dE/dx and the fact that the scattering cross-section is almost pure Rutherford have made the 1.0-MeV energy region the dominant one for materials applications of ion scattering.

In the constant dE/dx approximation the final energy, $E_1(t)$, of a particle at normal incidence that backscatters from depth t is given by

$$E_1(t) = K_{M_2}(\theta)\left[E_o-t(dE/dx)_{E_o}\right]+(t/\cos\theta)(dE/dx)_{E_1} \quad (4.1.17)$$

where the subscripts on dE/dx refer to the energy at which dE/dx is evaluated. This equation is simply an extension of Eq. (4.1.14) and uses the fact that at depth t the incident particles energy, E_t is

$$E_t = E_o - t(dE/dx)_{E_o} \quad (4.1.18)$$

A variation in thickness δt corresponds to a variation in energy δE_1 through

$$\delta E_1 = -\delta t\left[K_{M_2}(dE/dx)_{E_o} - (dE/dx)_{E_1}/\cos\theta\right] \quad (4.1.19)$$

so that the depth resolution in an experiment is determined by the energy resolution of the detector δE,dE/dx and geometry. Typical depth resolution is on the order of 3 to 10 nm.

Projectiles can scatter from any depth t, resulting in a continuous energy spectrum. The yield from a slice of width Δt at depth t is given by (for $\theta = 180^o$)

$$Y(t)\Delta t = [(Z_1Z_2e^2)/(4E_t)]^2 N\ Q\Delta\Omega\Delta t \qquad (4.1.20)$$

where N is the volume density of atoms/cubic centimeter in the solid. To convert this to a spectrum $N(E_1)dE_1$ of the measured energy E_1 we note that ($\theta = 180^o$)

$$dt = dE/[K(dE/dx)_{E_o} + (dE/dx)_{E_1}] \qquad (4.1.21)$$

and

$$E_t = (E_1+AE_o)/(K+A) \qquad (4.1.22)$$

where A depends on the ratio $(dE/dx)_{E_o}/(dE/dx)_{E_1}$ and kinematics. If $K\approx 1$ then $(dE/dx)_{E_1} \approx (dE/dx)_{E_o}$ and

$$N(E_1) \sim (E_o+E_1)^{-2} \qquad (4.1.23)$$

This spectral shape for E_o = 1.0 MeV is indicated in figure 4.3a

The scattering spectrum from a single crystal aligned with a major symmetry direction parallel to the beam is drastically modified from that of the amorphous target considered so far. In the amorphous target each atom senses a uniform distribution of impact parameters and obeys scattering laws as previously discussed The beam will undergo small-angle multiple scattering, in an uncorrelated fashion, which does not affect this impact parameter distribution. In an aligned single crystal the impact parameter distribution is also uniform at the first monolayer. However, the small angle scattering events determine the impact parameter for the second atom in a correlated fashion, and so on throughout the crystal. This correlation produces a unique (nonuniform) flux distribution inside the crystal which modifies the scattering probability. Since the flux distribution is known and the scattering probability measurable, this becomes a useful structure tool.

The simplest useful example of the effect of this small angle scattering is illustrated in the two atom model (Fig. 4.3b). Ions incident at the smallest impact parameter undergo large angle scattering, while those at large impact parameter suffer small deflections which determine the flux distribution of ions near the second atom. In a static representation this flux has a distance of closest approach to the second atom, R, within which there are no ions. For Coulomb scattering

$$R = R_c = 2(Z_1 Z_2 e^2 d/E)^{1/2} \tag{4.1.24}$$

where d is the atomic spacing and E is the incident ion energy. For 1.0 MeV H^+ incident along the <001> direction of W, R_c = 0.0164 nm. Since the bulk value of ρ, the two-dimensional thermal vibrational amplitude is 0.006 nm (at room temperature), the shadowing is still substantial, even in a nonstatic case.

In the Coulomb approximation the flux distribution at the second atom, $f_c(r)$ is given by |34|

$$f_c(r) = \begin{cases} 0, & r<R_c \\ (1/2)\left[(1-R_c^2/r^2)^{1/2} + (1-R_c^2/r^2)^{-1/2}\right], & r>R_c \end{cases} \tag{4.1.25}$$

This distribution is extremely sharp on the scale of finite vibration amplitudes and when folded with the Gaussian positional distribution of the atoms leads to an estimate of the two atom surface peak intensity in the Coulomb approximation as

$$I_c = 1 + \left[1+R_c^2/(2\rho^2)\right]\exp\left[-R_c^2/(2\rho^2)\right] \tag{4.1.26}$$

The first term represents the unit contribution from the first atom in the string, and the second term represents the variable contribution from the second atom. In most cases this two atom Coulomb approximation is not adequate for detailed comparison with experiments, however, it is useful as an illustrative example and for suggesting that the surface peak intensity is a function of one parameter, ρ/R. In cases of practical interest the surface peak yield corresponds to scattering from a few atoms per string. Furthermore, a screened Coulomb potential is necessary rather than the Coulomb approximation. Analytical approaches are then formidable and maybe even impossible. The most direct approach is computer simulation in which a large numer of incident ion trajectories are computed and the large angle scattering probability calculated for each layer, as outlined by Barrett |37|. An additional advantage of the simulation technique is that particular surface structure models are easily included for comparison with experiments.

The energetic ion beams of interest can penetrate microns into the bulk. When the beam is parallel to a major symmetry direction of the crystal the particles acquire channeling trajectories and, eventually, the channeling flux distribution. Channeling effects have been of considerable interest in recent years both as a solid state tool and for a variety of other purposes |38|.
For most of the applications described here the details of the bulk effects give primarily a qualitative description of the evolution of the channeling process. As described, the uniform incident ion beam encounters the first monolayer of the solid and undergoes small angle scattering to acquire the "shadow cone" flux distribution. (see Fig.4.3b). A correlated sequence of scatterings can continue for hundreds of atoms, until an incident particle is gently steered away from the closest string of atoms. Interaction with an adjacent row is similar, so that the initial motion is oscillatory in nature bouncing from string to string, with a wavelength hundreds of angstroms long. The important fact is that particles cannot get close enough to the atoms of the solid to undergo the very small impact parameter process such as large angle Rutherford scattering. Eventually the phases of the different trajectories are mixed and the flux distribution can be described by statical equilibrium concepts. Under this assumption an analytical treatment is possible. In the case of incidence precisely along the channeling direction the flux distribution, $f_B(r)$, well inside the crystal is of the form

$$f_B(r) = \ln[Nd/(Nd-\pi r^2)] \qquad (4.1.27)$$

where r is the perpendicular distance from the atomic row and N is the volume density of atoms inside the solid, so that Nd is the area of a unit cell. This is a distribution which is sharply peaked at the centre of the channel (i.e., far from the string of atoms) and decreases as r^2 to zero intensity at the atomic sites.

The scattering yield from bulk atoms is then the usual fold of the Gaussian positional distribution with the bulk flux distribution and is roughly given by $Nd\pi\rho^2$. Since the distribution is normalized to unity for nonchanneling conditions, the channeling bulk yield is the order of 1% of the nonchanneling yield. These results allow one to develop a description of the backscattering energy distribution in the channeling case. At energies corresponding to the surface of the crystal there is a surface peak corresponding to at least one monolayer. At lower energies, deeper depths, the scattering yield undergoes some oscillations and then reaches a steady state value of ~1% of the nonsingle-crystal yield: At even greater depths the dechanneling processes occur.

Let us now proceed with the investigations on the GaSb/AlSb superlattice |27|. The selection of the GaSb/AlSb system has the advantage of sharing the same anion, Sb, which enables the separation of the row offset model from the strain study. GaSb/AlSb (30nm/30 nm) periodic structures width 10 periods were grown epitaxial on (100) GaSb substrates by MBE. Channeling measurements and analysis were made using a 2-MeV Van de Graaff accelerator. Fig.4.4

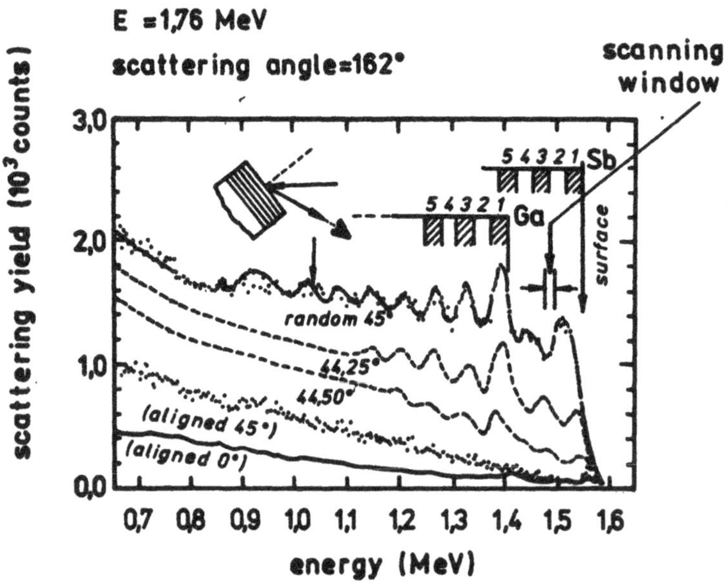

Fig.4.4 Energy spectra of 1.76-MeV $^4He^+$ ions backscattered from (100) GaSb/AlSb superlattices. Depth scale based on Sb signals and Ga signals are marked in the unit of number of layers (30 nm/per layer), [100] aligned, [110] aligned random, and two more spectra are given (after Chu et al., Ref.27).

shows a series of energy spectra of 1.76-MeV $^4He^+$ ions backscattered from a superlattice sample at various experimental conditions. The best channeled spectrum, i.e., the lowest curve in this figure is obtained with the ion beam aligned to the [100] growth direction (0° with respect to the surface normal) of the GaSb/AlSb superlattice. The next lowest spectrum, plotted as solid dots, is an aligned spectrum for [110] axial channeling (45° tilted from normal). A very broad energy window is set for the alignment to ensure that the alignment is based on a sampling of scattering from several layers inside the superlattice. The higher dechanneling along the [110] direction compared with that along the [100] growth direction for this superlattice sample is consistent with all the earlier measurements |25|,|26| on superlattices. The random spectrum was obtained by tilting the surface normal of the sample to 43° away from the beam direction with a 10° rotation away from the {110} plane. It is interesting to learn that the precise angular position for the best alignment along the <110> direction is a function of where the energy window is set. In Fig. 4.4 an energy window of about 15 keV [3 channels] is shown. In the present example this corresponds to a depth in the superlattice between the second and the third layers. Since the energy scale of this figure can be translated to a depth scale, a change in the energy window corresponds to probing different depths of the sample. In Fig. 4.4 the energy positions are indicated which correspond to the individual GaSb and AlSb layers in both the Sb part of the spectrum (1.4-1.6 MeV) and to those of the overlapping counts from Ga and Sb (below 1.4 MeV). Each layer corresponds to an alternating GaSb (1,3,5,...) and AlSb (2,4,6,...) layer, with a layer thickness of 30 nm.

Fig.4.5a shows the backscattering yield normalized to the random level (around 45° tilt) plotted as a function of the tilt angle. Two window settings are given, the first one is set between 1.535 to 1.550 MeV corresponding to the Sb signal of the surface layer of GaSb (plotted as solid circles). The second window is set between 1.490 to 1.505 MeV which corresponds to the Sb signal of the second layer of the superlattice. The centre of an angular yield curve is defined by the average of the two angular positions corresponding to the midheight on either side of the angular scan. It is believed that the centre corresponds to an angle of incidence giving the least dechanneling - "best" channeling for a given layer. One can see from Fig. 4.5a, that the two angular yield curves for two different depths do not have a common centre. The first layer is centred at 45.09°, and that of the second layer

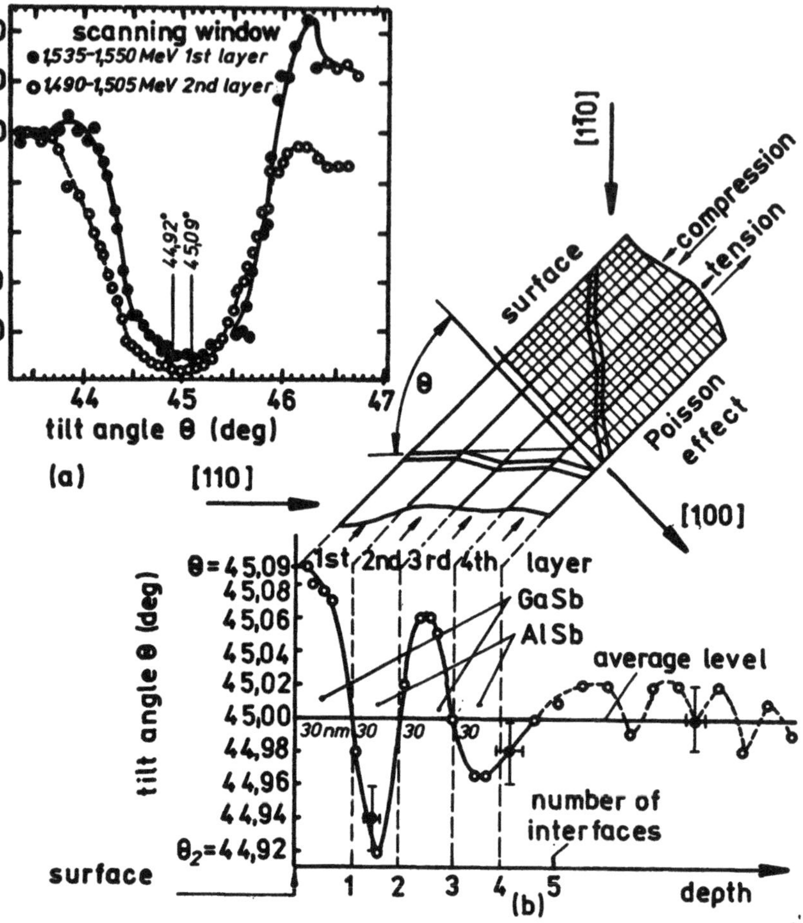

Fig.4.5. a)Angular scan by setting an energy (depth) window at the first layer (solid circles) and second layer (open circles) for 54 spectra run at 54 different angles. The centre position of the angular scan changes from layer to layer indicating the [110] direction variations
b)Top portion shows a schematic drawing of a strained superlattice and its influence on the channeling. The bottom part shows the angular position of the minimum yield defined by figure (a) as a function of depth. The oscillations are evidence of a kinked [110] channel.(after Chu et al.,Ref.27).

at 44.92°. This is a direct evidence that the [110] axis for the second layer is not in line with that of the first layer. A small angular difference of 0.17° is observed. This difference is presumably due to the lattice strain caused by intrinsic lattice mismatch between GaSb and AlSb in the superlattice.

These observations of the strain can be elaborated as is shown in Fig. 4.5b. The top part of this figure illustrates a model of a strained superlattice (Poisson effect), and its effect on the channeling along the [1$\bar{1}$0] or [110] direction. It has been shown that lattice mismatched heteroepitaxy can be grown with essentially no misfit defect if the layers are sufficiently thin and the mismatch is accommodated totally by uniform lattice strain |39|. The resulting strains in the superlattices consist of both hydrostatic and [100] uniaxial components which alter the lattice constants. For the discussed superlattice sample, the following relations are valid

$$a_o(AlSb)_{\perp} > a_o(AlSb) > a_{o_{\parallel}} > a_o(GaSb) > a_o(GaSb)_{\perp} \quad (4.1.28)$$

where $a_o(AlSb)$ and $a_o(GaSb)$ are bulk lattice constants, $a_{o_{\parallel}}$ is the lattice constant in the planes parallel to the interfaces, and $a_{o_{\perp}}$ are lattice constants for GaSb or AlSb perpendicular to the interfaces. Lattice strain cause the [1$\bar{1}$0] and [110] channeling direction to oscillate between an angle greater than 45° and an angle less than 45° degrees. This is what has been observed experimentally and is presented in the lower portion of Fig. 4.5b. In this figure the angular position of the minimum yield, as defined in Fig. 4.5a, is plotted as a function of depth of the superlattice. The window width is three channels wide equivalent to a 15-keV energy interval and is the same as the energy resolution of the backscattering system. From the energy-loss calculation, 30-nm GaSb produces a 34-keV energy shift, while 30-nm AlSb produces a 29-keV energy shift at the chosen experimental conditions. The oscillation of the angular position of the minimum yield is a direct evidence of the alternating tensile and compressive nature of the strain. The damping of the oscillations is due to the fact that any ion channeling at a given layer is always influenced by the previous history of the ion trajectory. We may assume that the change in angular direction of the [110] axis at each interface is the same. For all layers, the lower limit of this "kink" angle is given by the difference of the angular positions between the first and second layers, and from Fig. 4.5b this is found to be θ=.17° ± 0.03^{o}.

When the analysis reaches layer 5, i.e., the layer between interfaces 4 and 5, the backscattering yield information is complicated by crowding of the Sb signal from layer 5 and a portion of the Ga signal from layer 1. This can be seen from the overlapping of the depth scales of Sb and Ga given in Fig. 4.4. This restricts the considered depth analysis to the top four layers.

The channeling measurements of Fig. 4.5 indicate a ($0.17^\circ \pm 0.03^\circ$) "kink" between the layers of GaSb/AlSb. The results of the strain calculations are summarized in Table 4.1 based on the elastic constants of the materials. Because two different lattice parameters have been quoted in the literature for GaSb, the strain calculations are made for both values in separate columns of Tab.4.1. $a_{o\parallel}$ is related to the moduli and lattice constants of GaSb and AlSb. The value of $a_{o\perp}$ is calculated from the Poisson effect $\Delta x = -\Delta y C_{12}/C_{11}$ where C_{11} and C_{12} are moduli of elasticity. The kink angle is given by

$$\Delta\theta = \tan^{-1}[a_o(GaSb)_\perp / a_{o\parallel}] - \tan^{-1}[a_o(AlSb)_\perp / a_{o\parallel}] \qquad (4.1.29)$$

and is evaluated to be 0.37° or 0.15° depending on which lattice parameter is used for GaSb. The experimental value gives $0.17^\circ \pm 0.03^\circ$.

Table 4.1. Strain analysis for a GaSb/AlSb superlattice (after Chu et al.,Ref.27)

a_o(GaSb) nm	0.6095	Ref.40	0.6118	Ref.38
a_o(AlSb) nm	0.6135	Ref.40	0.6135	
$a_{o\parallel}$ nm	0.6115		0.6126	
$a_o(GaSb)_\perp$ nm	0.6077		0.6111	
$a_o(AlSb)_\perp$ nm	0.6155		0.6144	
$\Delta\theta$ calculated	0.37°		0.15°	
$\Delta\theta$ experimental		$0.17^\circ \pm 0.03^\circ$		

From the data presented above it is clear, that channeling can be used to measure the minute amount of alternating tensile and compressive strain in a strained superlattice. The experimental result is in good agreement with that calculated from the elasticity of materials. Oscillation of the [110] direction versus depth and the amount of angular deviation are consistent with the "strained-layer superlattice model" and with optical measurements performed on the GaSb/AlSb superlattices |41|,|42|.

Combining the $^4He^+$ ion-channeling measurements with X-ray rocking curve analysis for the structural investigation of the GaSb/AlSb superlattices, an extended set of experimental data has been gained |43|. These data allow one to formulate the following conclusions:

(i) X-ray rocking curve analysis is a very powerful method for characterizing the strain in superlattices as a function of depth, since it determines the geometry of the distortion (parallel, perpendicular, negative, or positive strain). The strain determined by this method is in excellent agreement with that calculated from the elastic constants and lattice misfit of the layer materials.

(ii) Backscattering and channeling measurements, combined with computer simulation, also allow a determination of the strain by providing a value of the associated "kink angle". The strain determined in this way is in reasonable agreement with the X-ray and elastic misfit values.

(iii) He-ion-beam bombardment at extended doses relieves a large portion of the strain of the AlSb/GaSb strained superlattice. This limits the beam dose that can be used in ion backscattering measurements without perturbing the specimen. On the other hand, this ion-beam perturbation might be utilized to modulate strain in superlattices in a controllable manner. This has potential processing applications in integrated optics involving superlattice layered structures.

4.2. Optical transitions in superlattices

Optical techniques have been shown to be a powerful tool for the investigation of superlattices |44|, for they provide direct information for the subband energies and electronic properties |45-54|, as well as for the phonon properties |55-60| of different superlattice structures.

4.2.1. Absorption and luminescence

In the compositional superlattices the potential acting on the charge carriers is the sum of the potentials in the A and B materials and of a square wave shaped superpotential which arises from the electron affinity and band gap difference between A and B. For the conduction band for example, the layers of one material-

-say A-form a series of square shape potential wells which confine the conduction states. To each of the quantized energy levels related to the conduction states corresponds a two-dimensional density of states $m_e^*/\pi\hbar^2$. The associated wave functions are mainly localized in the A layers, with evanescent wings in the adjacent B layers. These potential wells are separated by the potential barriers formed by the layers of the B material. Tunneling of electrons through these potential barriers leads to the formation of subbands having small but finite widths, which are the superlattice (SL) conduction bands. Thus, superlattices present a duality of aspects, the quantum well aspect with its associated notions of quantized confinement energies, two-dimensional densities of states and wave function localization, and the superlattice aspect, with the strong anisotropy of the Brillouin zones and the resulting smallness of the bandwidths in the z-direction. As for the superpotential associated with the valence band, the effective mass dependence of the quantized confinement energies [see, Eq. (3.2.2)] shows that the degeneracy of the host Γ_8 valence band is lifted in the SL, the heavy-hole states having smaller confinement energies than the light hole states. Indeed, a more careful analysis |44| shows that the |3/2, ±3/2> (heavy holes) and |3/2, ±1/2> (light holes) states are strictly uncoupled at $\vec{k}_{\parallel} = (k_x, k_y) = 0$. This is no longer true for finite $\vec{k}_{\parallel}$.

Up to now, most of the optical studies in type I superlattices have been performed on MQW structures, for which the superlattice aspect is of secondary importance. On the other hand, the type II InAs/GaSb superlattices generally display conduction subbands widths of a few tens of meV, and they present some three-dimensional character.

One of the most striking pieces of evidence of the quantization of the energy levels in $Al_xGa_{1-x}As$/GaAs MQWs was obtained from optical absorption measurements |61-63|. The transmission spectra exhibit a series of sharp exciton peaks separated by absorption plateaus characteristic of the constant two-dimensional density of states (see, Fig.4.6a). For very thin (<20 nm) layers, the first exciton peak is clearly split, which results from the difference between the heavy and light hole confinement energies. From fits of such data, the offsets of the hosts bands-edges ΔE_C and ΔE_V can be determined. From the assignment of the 2S exciton absorption,

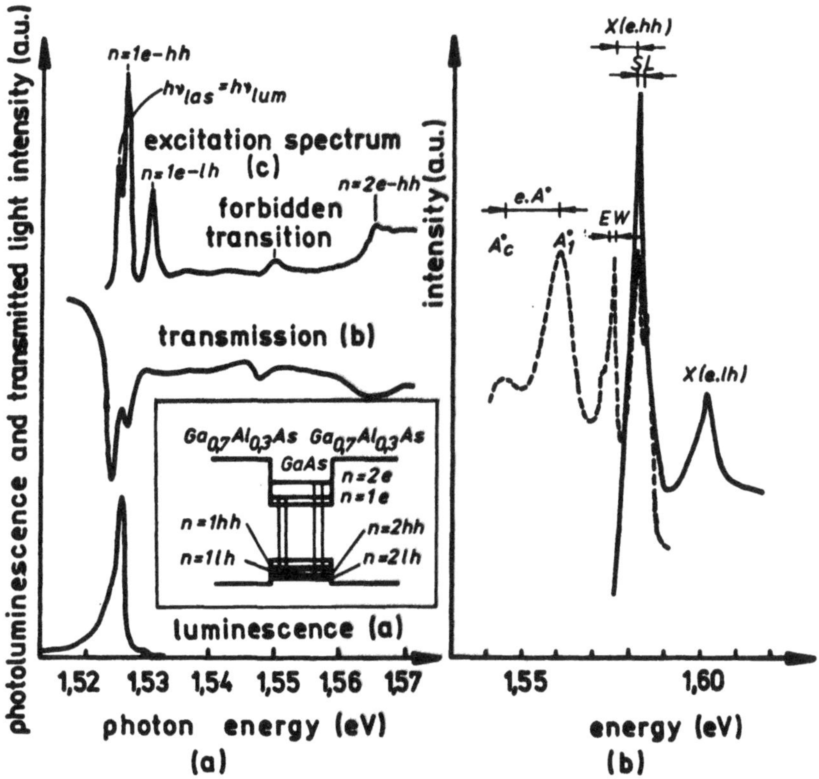

Fig.4.6. a) Photoluminescence, absorption, and excitation spectra in a 18.8 nm thick GaAs quantum wells of a 25 period $GaAs/Al_{0.3}Ga_{0.7}As$ MQW structure, measured at T=1.8 K. The various levels are identified in the inset (after Weisbuch et al.,Ref.62).
b) Photoluminescence spectrum dashed line of a 7x7 nm $GaAs/Al_xGa_{1-x}As$ superlattice measured at 4.2K, and the photoluminescence excitation spectrum of the same sample (after Deveaud et al.,Ref.65).

it has been concluded |44|, that the "light" hole exciton should have a slightly larger binding energy than the "heavy" hole exciton. Finally, the reported figures for the absorption coefficient |61| are in qualitative agreement with the calculated values. Moreover, the parity selection rule (see,Sec.3.1.2) seems fully obeyed since only weak $\Delta n=2$ transitions and no symmetry forbidden transitions, are observed. Similar information on the shape of the absorption coefficient is obtained from excitation spectroscopy. This technique was used to determine the origin of the excitonic absorption linewidth. From the experimental data of Weisbuch et al. |64| it has been concluded that this linewidth is governed by intralayer thickness fluctuations which appear to be essentially one monolayer in height and a few hundreds of angstroms in lateral size.

This conclusion has been confirmed by an extended luminescence study performed by Deveaud et al. |65|. The samples they studied were either regular GaAs/AlGaAs superlattices with well width L_z ranging between 7 and 4 nm and barrier width L_b equal to L_z, or MQW heterostructures with much thicker barrier layers, or SLs perturbed by the introduction of enlarged wells (EWs). A typical luminescence spectrum is presented in Fig. 4.6b (dashed line). Three main structures are observed for a SL structure with L_z = 6.6 nm and L_b = 6.3 nm:

(i) the exciton peaks of the regular SL at 1.58 eV,

(ii) a peak corresponding to the excitonic recombination in the EWs at 1.575 eV, and

(iii) a band peaking at 1.560 and 1.544 eV corresponding to electron acceptor transition with the acceptor located near the interface or at the centre of the wells.

These attributions can be checked by photoluminescence excitation (PLE) experiments. When detection is set on the 1.584 eV peak, two transitions are observed at 1.584 eV and 1.603 eV, respectively corresponding to heavy-hole and light-hole SL exciton transitions (see, Fig. 4.6b solid line). If detection is set on the EW peak, PLE shows four peaks as a result of carrier motion in the growth direction |66|. As can be seen in Fig. 4.6b, both the SL exciton peak and the EW peak are split. Furthermore, it must be stressed that the linewidths are smaller than both the predicted splitting for a monolayer well width fluctuation and the third of the distance between the SL and EW excitonic peaks. In any case, low excitation densities favour the low energy component of each peak. This splitting is observed in any SL or MQW grown in the optimized conditions, with well width between 4 and 7 nm, and with or without EWs introduced in the structure. For the SL peaks, the value of the splitting corresponds to the calculated energy shift introduced by a single monolayer difference in the well width (0.285 nm). If for any reason (substrate heater disfunction, et...), the optimized conditions (690°C as substrate temperature, V to III flux ratio close to 3) are not fulfilled, the linewidth becomes larger and the splitting is not observed.

Interpretation of these results is straightforward using the model introduced by Weisbuch et al. |64| for interface disorder. This model is shown in Fig. 4.7a, for a case corresponding to a

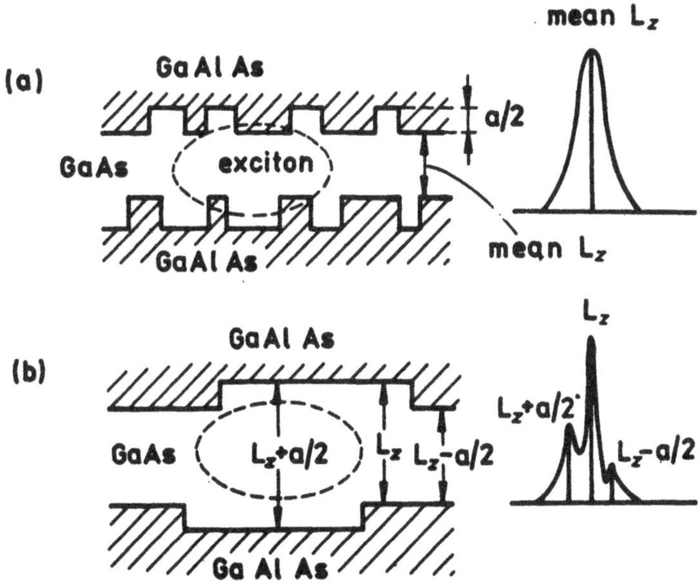

Fig.4.7. a) Effect of the interface disorder on the luminescence linewidth when the lateral size L_S of the growth islands is smaller than the exciton Bohr diameter. The line with corresponds to the confinement energy fluctuations experienced by the exciton.
b) When L_S is larger than the exciton Bohr diameter, the exciton only experiences one confinement energy corresponding to L_Z-a_o/2,L_Z, or L_Z+a_o/2 and discrete peaks are observed in luminescence (after Deveaud et al.,Ref.65).

good growth quality. In such a case, growth islands of one monolayer thickness will be present at each interface, so that the well width will show three possible widths: the nominal width L_z and L_z+ a_o/2 or L_z- a_o/2. Growth quality will be further qualified by the lateral size L_S of these growth islands: the higher the growth quality the larger this size will be. If L_S is smaller than the exciton Bohr diameter (case a), the PL and PLE peaks will show a linewidth corresponding typically to the variation in confinement energy due to $a_0/2$ fluctuations. This is due to the fact that the exciton simultaneously experiences different well width If, on the contrary, L_S is larger than the exciton Bohr diameter (case b), an exciton will only experience one well width at a time. Thus three different exciton peaks will appear, in place of the only one peak of case a, corresponding to the three widths seen by excitons. The linewidth of each of these peaks is then smaller than in case a. As motion in the x-y plane is easy, thermalization might be expected to occur, at least in part, and thus the larger well size $L_z+a_0/2$ or the smaller confinement energy should be favoured.

The splitting of SLs exciton peaks in the considered samples may therefore, be interpreted as originating in single monolayer fluctuations in each well. It is possible to imagine that, rather than intralayer size fluctuations, some variations of the mean well width from one layer to another are observed. The first argument against this assumption is that there is no reason why such variations should be step-like, giving rise to discrete peaks and not continuous. Each well would have its own mean width and the total result would be a broadening of the luminescence peaks. Furthermore, the growth time is controlled with an accuracy better than 50 msec, the growth rate being 0.2 nm/sec. Finally, it cannot be assumed that, using the spatial resolution of 50 μm spot, no monoatomic step is present at each interface, so that one necessarily has at least three possible well thicknesses. It might be expected to observe no discrete peaks, but in that case, the linewidth would have to be larger than the confinement energy fluctuations due to one monolayer width fluctuations. The extrinsic nature of the peaks is ruled out by PLE experiments: the splitting of the SL excitons is also observed in PLE, both on the heavy-hole and on the light-hole excitons. Positions of the heavy-hole excitons exactly coincide with the two peaks observed in luminescence. The low energy PLE peak is favoured both for heavy-hole and for light-hole excitons when detection is set on the low energy luminescence peak, the high energy peaks being favoured for detection on the high energy luminescence peak.

Due to the position of the ovens and to the fact that in the MBE growth no rotating sample holder has been used |65|, the SL parameters vary along one of the sample directions (say x). A typical variation is two monolayers over the sample length (1 cm) for a 15 nm SL period. This variation is observed by X-ray diffraction. Variations of the luminescence peaks positions along x confirm this variation. In good quality samples, the modifications of the shape of luminescence peaks are in agreement with the discrete values of L_z and with the changes in the proportions of $L_z-a_o/2$, L_z and $L_z+a_o/2$ well widths. As a consequence, at some places of the same sample, two, or three peaks are observed (see, Fig. 4.6b and Fig. 4.7b). When the spot position changes slightly the position of the individual peaks almost stays constant while the mean position and so the intensities of each component change. For large movements of the spot, some components disappear and are replaced by other peaks shifted by an energy almost corresponding to an integer number of monolayers.

The type of splitting that is shown in Figs. 4.6 and 4.7, can be expected as long as the exciton is located in one well only. When the SL period decreases, the exciton will have more and more 3D character and can be expected to extend over several wells. As the growth islands are not correlated from one interface to the next one, when the exciton will experience more than one well, it will no longer experience discrete confinement energies. As a matter of fact, in 3x3 nm SLs, even with very good luminescence peak width (3 meV, while the monolayer splitting is 4 meV) discrete peaks are not observed.

Optical absorption measurements were also performed in InAs/GaSb superlattices grown by MBE |67 - 69| The very small absorption coefficients and the spectral position of the absorption edges clearly revealed the type II nature of these SLs and also the rather unusual relative position of the host band-edges. Indeed, the interpretation of the experimental data shown in Fig. 4.8a, implies that the InAs conduction band lies about 150 meV below the top of the GaSb valence band, as predicted from the comparison of the electron affinities. Two major consequences result from this situation: (i) the InAs Γ_6 conduction band always strongly interacts with the GaSb Γ_8 light-hole band, which is the reason for the "large" bandwidths in these SLs. (ii) The SL band gap [defined as the energy difference between the first conduction (E_1)

and heavy-hole (HH_1) subbands], which decreases with increasing layers thicknesses, actually reaches negative values when the sum of the electron and heavy-hole confinement energies become smaller than 150 meV, which is the case for periodicities larger than 18 nm and equal layer thicknesses. The extrapolation of interband magnetooptical transitions towards negative energies gave direct evidence of this configuration |44|. The absorption spectra of semiconducting InAs/GaSb SLs shown in Fig. 4.8a exhibit step-like

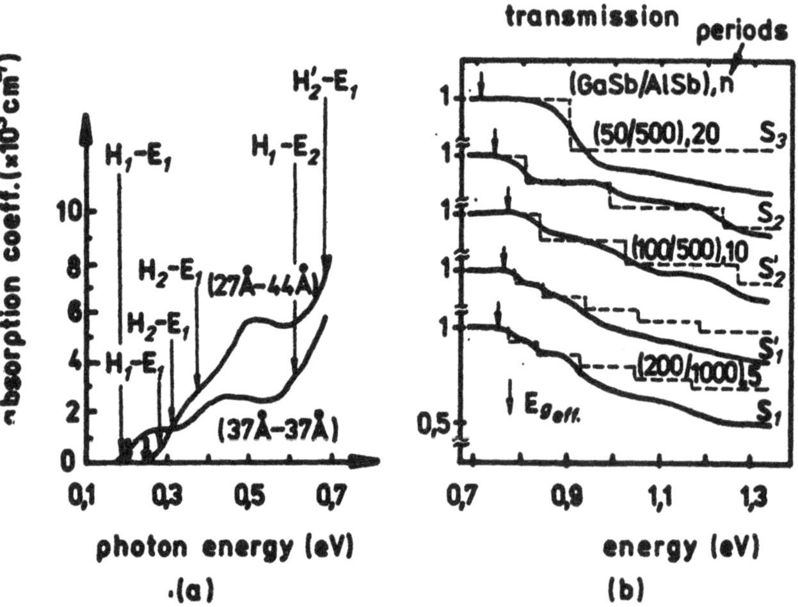

Fig.4.8. a) Absorption spectra of two InAs/GaSb compositional superlattices (after Chang et al., Ref.68)
b) Transmission spectra of GaSb/AlSb MQWs. The samples S_1 and S_2 belong to a different series of growth (after Voisin et al.,Ref.41).

structures which are attributed to transitions between the various valence and conduction subbands. In agreement with the selection rules of section 3.1.2, transitions with $\Delta n=0$ or $\Delta n\neq0$, and Δn even or odd present equivalent strengths. Moreover, the reported absorption coefficients are in good agreement with detailed calculations |44|. The absence of exciton peaks in these spectra should not be regarded as an indication of poor sample quality. In fact no excitonic effect should be expected in these structures, as the spatial separation of the carriers in type II SL result in a drastic reduction of the binding energy of an electron-hole pair SLs with good crystalline quality can also be grown from materials with a rather large lattice mismatch |39|. Photocurrent spectroscopy was used to determine the energy gap of various SLSs |70|, but the reported data can hardly establish the two-dimensional character of these structures. The transmission spectra shown in Fig. 4.8b, which have been obtained by measuring the optical absorption at low temperature in MBE grown GaSb/AlSb MQWs where $\Delta a_o/a_o = 0.65\%$, exhibit the step-like behaviour characteristic of a two-dimensional density of states. From the series of absorption steps it is clear that the involved conduction band states are those associated with quantized electron states in the GaSb layers. The intriguing feature is that the absorption edges are observed at energies significantly smaller than those expected for simple GaSb quantum wells. However, the type I nature of this system is definitely established by the consideration of the order of magnitude of the absorption coefficients. Similar results have been obtained when optical absorption was measured on the $In_xGa_{1-x}As$/GaAs strained MQW structures |52|. However, the first observation of the quantum size effect in strained-layer superlattices has been reported for the $GaAs/GaAs_xP_{1-x}$ SLS |51|. Using excitation and photoluminescence spectroscopies, optical transitions between excited states of the electron and hole quantum wells were observed in this SLS. These observations provide a measure of the biaxial compressive strain in the GaAs layers, interfacial roughness and valence-band offset for the GaAs/GaP heterojunction.

Superlattices of HgTe-CdTe have been proposed as a novel material for applications as infrared materials. These man-made systems have the advantage of allowing one to adjust a number of the properties of the material so that the interesting features are near optimum for a given application in the infrared |71|,|72|.

The first calculations of the band structure of HgTe-CdTe superlattices were published some years ago |73|. Since that time some other band structure calculations for these superlattices have been done |71|,|74-76|. Then the effect of an increase of the band gap of the superlattice due to Hg interdiffusion has been theoretically predicted |77|. The energy position and width of the first conduction (E_1), heavy-hole (HH_1), and light-hole (h_1) subbands calculated at 4 K as a function of layer thickness, and the temperature dependence of the band gap $E_g=E_1-HH_1$ of the HgTe-CdTe SL |76| are shown in Fig. 4.9.

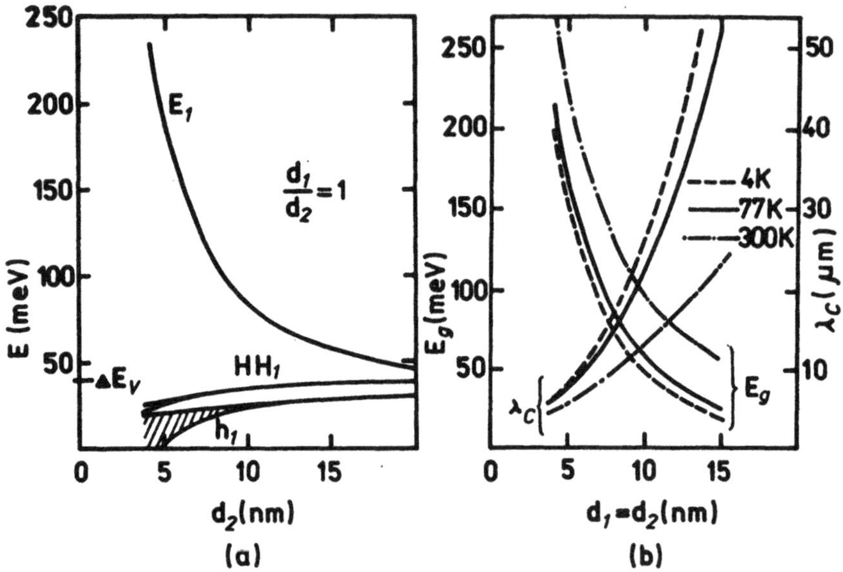

Fig.4.9. a) Energy position and width of the first conduction (E_1), heavy-hole (HH_1), and light-hole (h_1) subbands calculated at 4 K as a function of layer thickness for HgTe-CdTe SLs with $d_1=d_2$, where d_1 and d_2 are the thicknesses of the HgTe and CdTe layers, respectively.
b) Variation of the band gap E_g and of the corresponding cut-off wavelength λ_c calculated at 4,77, and 300 K as a function of layer thickness for the same SLs (after Guldner et al.,Ref.76)

It should be noted that there are two important parameters in these superlattices: d_1 the thickness of the HgTe well layers which controls essentially the superlattices band gap, and d_2 the thickness of the CdTe barrier layers which governs the width of the superlattice subbands, i.e., the electron effective mass along the superlattice axis. This is to be compared to the situation in bulk $Hg_{1-x}Cd_xTe$ alloys |72|,|75|, where only one parameter (namely x) is adjustable.

First experimental data related to the band structure of the HgTe-CdTe SLs have been gained by using the far-infrared magnetoabsorption technique |50|,|78|. These measurements performed at 1.6 K have shown that these SLs with geometrical parameters d_1=18 nm and d_2=4.4 nm are quasi-zero-gap semiconductors, displaying electronic properties which are found neither in bulk HgTe and CdTe crystals nor in $Hg_{1-x}Cd_xTe$ ternary alloy. For example, at low magnetic fields the electron effective mass in the plane of the layers is strongly nonparabolic and much lighter than in bulk HgTe. The theoretical calculations reported in the references listed above, predicted that the interesting cut-off wavelengths for infrared detectors, i.e., 8-12 µm |72| could be obtained for d_1 values of the HgTe-CdTe SL laying in the range of 5-7 nm. This prediction has been confirmed experimentally for 300 K, by performing measurements of optical absorption for three SLs whose characteristics are listed in Tab. 4.2.

Table 4.2. Characteristics of superlattices HgTe-CdTe grown at 185°C. n is the number of periods and $d_{1,2}$ are the thicknesses of each kind of layers (after Faurie et al.,Ref.79)

Superlattice sample	n	d_1 (nm) HgTe	d_2 (nm) CdTe	λ_c µm
1	150	4.	6	2
2	91	8	4	5.5
3	102	10	4	7

The infrared transmission curves are not very sharp due to the small total thickness of the SLs: 1-1.5 µm. Thus, the onset of the absorption and the half maximum of this absorption should be considered. The positions of these two points are shown in Fig. 4.10a, for the three SLs as a function of the HgTe layer thickness. In this figure also the calculated cut-off wavelength curve at 300 K, and d_1=d_2 |76| is shown for comparison. The experimental results confirm: (i) the increase of SL λ_c as compared to the alloy λ_c with relevant composition x at 300 K (see,Fig.4.10b), and (ii) the increase of λ_c as HgTe layer thickness increases.
The agreement at 300 K between theory and experiment is good if one considers the onset of the absorption. It has been shown by these experiments, that the interdiffusion between HgTe and CdTe

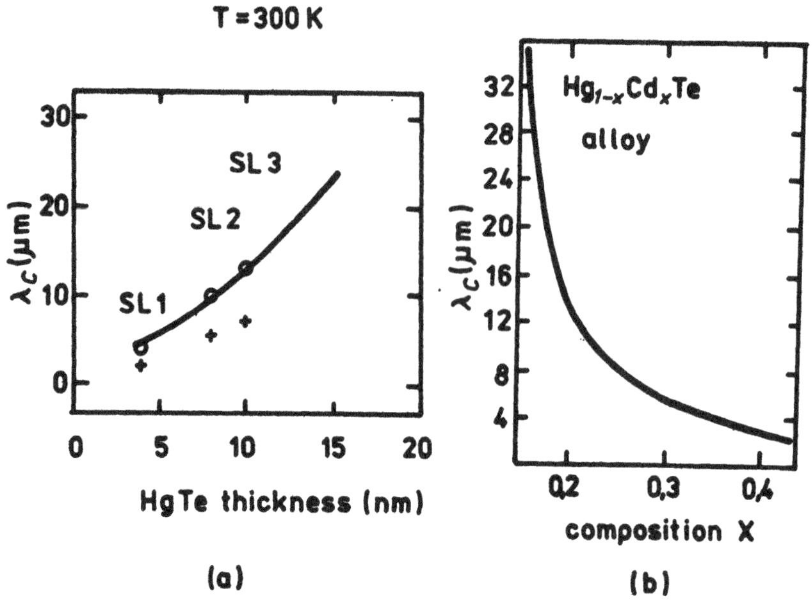

Fig.4.10 a) Calculated cut-off wavelength λ_c for a HgTe-CdTe superlattice of equal layer thicknesses $d_1 = d_2$, as a function of d_1 (solid line). The circles and crosses represent the experimental wavelengths of the onset of the optical absorption and the half maximum of the absorption, i.e., the conventional λ_c, respectively (after Faurie et al.,Ref.79)
b) Cut-off wavelength as a function of the $Hg_{1-x}Cd_xTe$ alloy composition x (after Smith et al.,Ref.75).

should be in the 1 nm range. This interdiffused region is a kind of alloy with a non-constant composition whose absorption is non--sharp but which occurs in a shorter wavelength range than the SL This absorption due to the interdiffusion cannot be neglected in HgTe-CdTe superlattices and could explain the absorption tail λ_c shifted towards shorter wavelenghts. The experimental results are related to a HgTe-CdTe SL presenting an absorption region in the far infrared window 8-12 μm. When the optical absorption experiments are carried out at 30 K, no important change for all superlattices investigated is observed, in opposition to theoretical predictions |76|.

One of the most interesting optical properties observed in diluted magnetic semiconductor (DMS) superlattices is the fact that the photoluminescence from these structures is strikingly more intense compared to the numerous thin film and bulk samples of CdTe and $Cd_{1-x}Mn_xTe$ that have been examined in the past |80|. When comparing photoluminescence spectra obtained for a superlattice consisting of alternating layers of 30% and 8% Mn, and for a 1.6 μm film with 8% Mn grown under identical conditions as the 8% Mn layers of the superlattice then one may realize that the superlattice peak is 1500 times more intense than the band-edge emission from the film under identical conditions of excitation. Moreover, since the FWHM of the superlattice spectrum is 3.8 times larger than that of the film, the quantum efficiency of the superlattice appears to be 5700 times greater. When comparing the photoluminescence intensity of various superlattices with a series of thin films and bulk samples, the SLs consistently show about three orders of magnitude enhancement in photoluminescence. It should be noted that the MBE films of $Cd_{1-x}Mn_xTe$ used in the investigations were of exceptionally high quality, as indicated by the FWHM of 8 meV. This value is less than the 11 meV typical of bulk samples of comparable Mn content. It is also less than the previously reported values for MBE grown CdTe films |80|.

A comparison was also made between the 8% /30% $Cd_{1-x}Mn_xTe$ SL and an AlGaAs/GaAs double heterostructure grown by LPE. The III-V compound structure was composed of $Al_{0.31}Ga_{0.69}As/GaAs/Al_{0.31}Ga_{0.69}As$ deposited on an n^+-GaAs substrate, with layer thicknesses of 0.4,0.6, and 0.8 μm, respectively. The threshold current for injection lasers made from this structure has been found to be 1.5-2.5 kA/cm^2. As shown in Fig. 4.11, the DMS SL structure indicated a somewhat greater photoluminescence intensity, when compared to the AlGaAs/GaAs double heterostructure. The FWHM values for the superlattice (ternary) and double heterostructure (binary) are 31 and 10 meV, respectively. Petroff et al. |81| and Masselink et al. |82| have observed similar photoluminescence enhancements in III-V superlattices, and attribute the enhancement to a gettering of non-radiative centres at the interfacial regions. However, the possible and more important factor in the enhancement of photoluminescence is the exciton confinement in the quantum wells of the SL structure |83|.

Finally, photoluminescence measurements in DMS SLs carried out in the presence of a magnetic field show a large frequency shift of

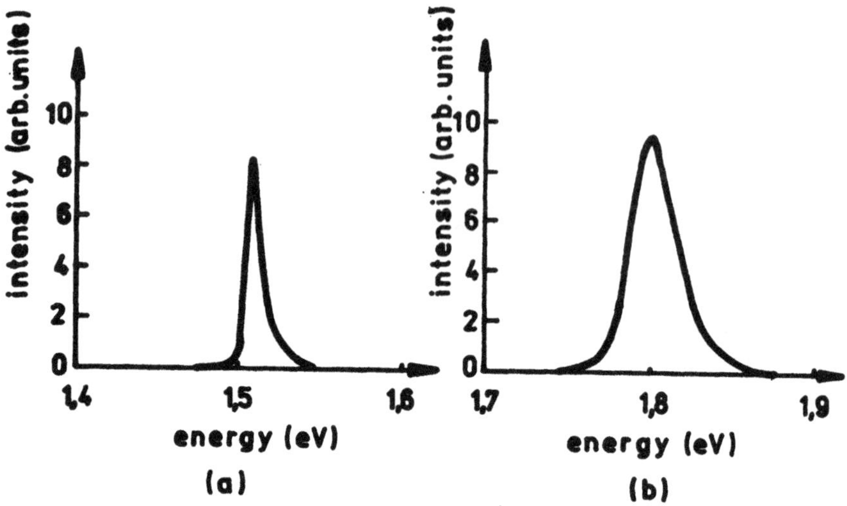

Fig.4.11. Comparison of photoluminescence for an AlGaAs/GaAs double heterostructure (a), and for a $Cd_{1-x}Mn_xTe$ SL observed under identical conditions (b). The vertical scale is the same for both cases (after Datta et al., Ref.80).

the photoluminescence peak, the value of which is highly temperature sensitive. For example, the $Cd_{1-x}Mn_xTe$ SL consisting of alternating 11% and 50% of Mn layers shows a shift of the luminescence peak at 5 K that corresponds to an exchange-induced effective g-factor of about 100. This demonstrates that the very large electronic g-factor for the $Cd_{1-x}Mn_xTe$ alloys are also realized in DMS SLs, so that the exchange-induced effects which have led to so many new and interesting phenomena in bulk DMS can be expected to play an important role in the behaviour of DMS SLs as well |80|.

4.2.2. Raman scattering

Raman scattering of light is a natural probe for superlattice phonon studies. It has been most frequently employed for studying the vibrational properties of these structures |55-60|. Moreover, Raman data can provide information on the layering parameters heretofore gathered primarily from X-ray diffraction studies |58|.

Several different approaches to modeling superlattice phonons have been found to be useful, each having certain advantages. A linear-chain model embodies the discrete nature of the crystal structure and provides information about evanescent behaviour of optical phonons. It is most useful in frequency regions in which the bulk-phonon dispersion is nonlinear. At low frequencies and long wavelengths the lattice can be treated as an elastic continuum, in which case a Kronig-Penney type of approach yields a convenient analytic expression for phonon folding. The layer thicknesses in this case are not restricted to integral multiples of the monolayer spacing. Alternatively, at these same frequencies, the superlattice can be considered as a perturbation on an average elastic background. This approximation approach yields explicitly the gap widths and sound velocity for an arbitrary composition profile. In addition to these models also a dielectric continuum model, which allows an estimate of the effects of long-wavelength electric fields, may be employed. Of course, one wants a comprehensive theory of superlattice effects which correctly deals with all microscopic short- and long-range forces. Such a theory is especially needed in dealing with the spectrum of optical phonons. Calculations using such a theory have been made by Yip and Chang |84|.

When performing Raman scattering experiments, the SL samples mounted in a He cryostat and held at temperatures between 2 and 300 K are illuminated with light from a CW Ar^+, Kr^+ or dye laser. The scattered light is usually analysed by a double-pass spectrometer equipped with precise, preferable holographic 1800 groves/mm, gratings and subsequently detected by a suitable photomultiplier tube operated in a photon-counting mode. The measurements are normally made with light incident at an angle close to Brewster's angle, polarized in the plane of incidence. The scattered light is collected with a lens, whose axis is normal to the surface of the SL sample. Successive rotations of a polarizer in the scattered beam allow polarized and depolarized spectra to be interspersed in the same spectral scan, so that small frequency shifts between the two spectra can be detected.

Let us now consider a specific case of Raman scattering investigations performed on a variety of $Al_xGa_{1-x}As/Al_yGa_{1-y}As$ SLs grown by MBE on (001)- oriented GaAs substrates |58|. In these investigations the following results have been found:

(i) Folded acoustic phonons appear as doublets in the Raman spec-

tra of the acoustic region (5-220 cm^{-1}). Their frequencies may be accurately predicted by several SL models, including an approximate solution of an elastic continuum model through a perturbation approach. Scattering intensities of the folded acoustic modes can be correctly predicted by a photoelastic continuum model.
(ii) Anisotropy of the optical phonons (optic region 220-420 cm^{-1}) can easily be studied by using the layered dielectric continuum model. Due to the two-mode behaviour of the $Al_xGa_{1-x}As$ alloys, the effect of layering on phonons near the GaAs optical frequency in these SLs depends strongly upon the value of x in the barrier layers. Linear-chain model calculations indicate that optical phonons in binary SLs (GaAs/AlAs) are largely confined to alternate layers. Peaks in the Raman spectra may be clearly identified with the resulting quantized optical modes.
(iii) Raman scattering investigations have the potential to provide structural information similar to that which can be obtained by X-ray diffraction measurements.
In the Raman scattering studies performed on Ge_xSi_{1-x}/Si strained-layer superlattices |57|, the Raman spectrum from these SLSs has been compared to that from commensurate and incommensurate single layers of the same composition. The frequency and half width of the Raman spectral lines give information about strain, dislocation densities and disorder in these materials. The experimental results confirmed that the lattice mismatch in commensurate single layers and SLSs is entirely accommodated by a homogeneous tetragonal compression of the alloy layers only.

4.2.3. Luminescence in doping superlattices

The absorption of photons in nipi-crystals has already been discussed in Sec. 3.1.2, therefore, here we can concentrate the considerations solely on the luminescence phenomena. These are strongly affected by the spatial separation between electrons and holes. As a consequence, the shape and the energetic position of the luminescence spectrum, as well as the luminescence intensity, are tunable quantities.

The excited carrier lifetime enhancement, caused by the spatial separation between electrons and holes, depends exponentially, first, on the design parameters of the doping superlattice |86|,

|87| and second, on the degree of excitation, i.e., on the actual effective gap $E_g^{eff,n}$ or quasi-Fermi level difference $\Delta E_{F_{np}}$. Therefore, doping SLs with high doping concentrations are of particular interest for luminescence studies, as rather high electron-hole recombination rates occur even at effective gap values far below E_g of the host material.

In Sec. 3.1.1 it was stated that thermalized holes normally populate an acceptor impurity band. This implies that there is no conservation of momentum parallel to the layers effective in the luminescent recombination between electrons in conduction subbands and holes in this acceptor impurity band. Thus, a step-like shape of the luminescence spectra is expected, which reflects the sum of the contributions of the various populated subbands with the appropriate recombination probabilities (see,Fig. 4.12)

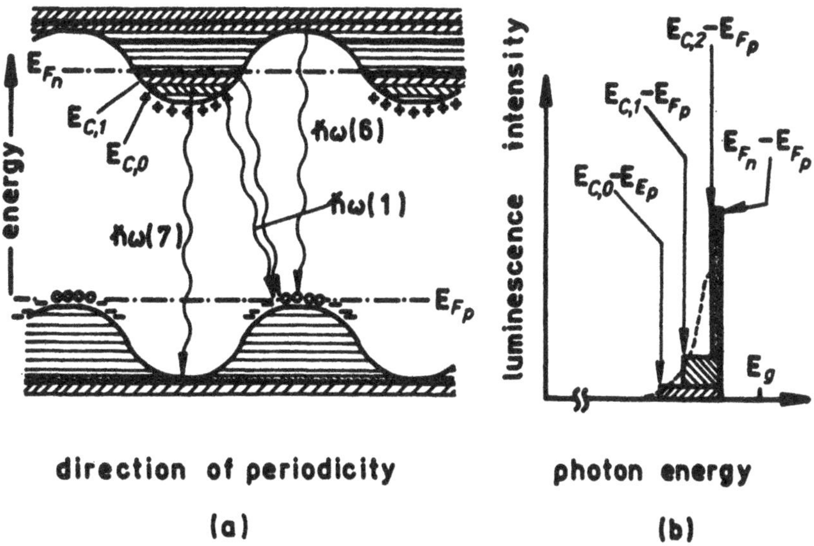

Fig.4.12. Schematic illustration of the luminescent processes in a GaAs doping superlattice (a) and the luminescence spectrum expected for such a SL with three electronic subbands populated (b) (after Ploog et al.,Ref.86)

The luminescence intensity at the photon frequency emitted per layer and per unit area is given by

$$I(\omega) \approx [4^2 \omega n_{opt}/(3\hbar c^3 m_o^2)] \int_{E_{c,0}}^{E_{F_n}} dE \sum_j \sum_\mu |\langle c,\mu|\vec{p}|a_j\rangle|^2 N^{(2)}_{c,\mu}(E) \cdot$$

$$\cdot \delta[E-(E_{F_p} + \hbar\omega)] \qquad (4.2.1)$$

if no broadening effects are taken into account (n_{opt} is the refractive index). The two-dimensional density of states $N^{(2)}_{c,\mu} = m_e^*/(\pi\hbar^2)$ is the same for all subbands above the respective subbands edge, i.e., for $E>E_{c,\mu}$. The square absolute value of the dipole matrix elements between conduction subband states and neutral acceptors at the sites $\vec{R}_j$ in the p layer $|\langle c,\mu|\vec{p}|a_j^o\rangle|^2$ increases strongly with the subband index μ. The recombination occurs between the spatially separated electrons near the bottom of the conduction subbands and holes in the acceptor impurity band across the indirect gap in real space. This process, described by Eq. (4.2.1), is indicated in Fig. 4.12a, by the arrows labelled $\hbar\omega^{(1)}$. During photoexcitation and at higher temperatures "vertical" transitions in real space, according to the processes discussed in connection with Eqs. (4.2.6) and (4.2.7) and labelled $\hbar\omega^{(6)}$ and $\hbar\omega^{(7)}$, respectively, are also possible.

The high-energy edge of the luminescence spectrum corresponds to the quasi-Fermi level difference

$$\hbar\omega_{max} = \Delta E_{F_{np}} = E_{F_n} - E_{F_p} \qquad (4.2.2)$$

At the photon energies

$$\hbar\omega_\mu = E_{c,\mu} - E_{F_p} < \Delta E_{F_{np}} \qquad (4.2.3)$$

the luminescence intensity decreases abruptly within this simplified model. It is obvious that the energetic position and the shape of the whole luminescence spectrum changes strongly as a function of the carrier concentration (see,Fig.3.10a). Also associated with the shift of the luminescence spectrum is a strong variation of the intensity since the magnitude of the momentum matrix elements decreases on an exponential scale if $\Delta E_{F_{np}}$ is lowered. As a consequence, the luminescence intensity of a doping superlattice may change by several orders of magnitude while the carrier concentration is changed only by a relatively small factor.

In Fig.4.12b, the expected luminescence spectrum corresponding to three populated subbands is shown schematically. The solid line corresponds to a situation with no line broadening taken into account. An asymmetric broad emission band (dashed line) results if strong broadening mechanisms are present. The step-like structure tends to disappear if a finite width of the acceptor impurity band is taken into account. Nevertheless, the shape of the luminescence spectrum, in particular its half-width, changes considerably and in a non-monotonous way if the population of subbands is varied. This is a consequence of the strong variation of Fermi energy $E_{F_n}-E_{c,\mu}$ in the uppermost populated conduction band when the carrier concentration is varied (see, Fig.3.10a). The largest width of the luminescence spectrum is expected for an electron concentration just below the threshold for population of the first excited subband. It is interesting to note that this situation may correspond to a quite low excitation intensity.

So far, only the luminescent recombination between relaxed electrons with energies below the electron quasi-Fermi level E_{F_n} and relaxed holes with energies above the hole quasi-Fermi level, E_{F_p}, have been discussed. At finite temperatures and/or during optical excitation one also has to take into account contributions to the luminescence due to radiative recombinations of electrons above E_{F_n} and of holes below E_{F_p} |87-89|. The relative importance of such processes may be enhanced in doping superlattices as compared to unmodulated bulk semiconductors. In both cases the population probability is small at electron energies

$$E_e-E_{F_n} \gg kT \qquad (4.2.4)$$

and hole energies

$$E_{F_p}-E_h \gg kT \qquad (4.2.5)$$

either because of the fast relaxation times (typically of the order of picoseconds or less) or because of the small values of the Fermi distribution function in the case of quasiequilibrium. The weight of contributions from electrons in "quasicontinuum conduction subbands", i.e., in the subbands at energies

$$E_e>E_{F_p}+E_g-E_A \qquad (4.2.6)$$

or of holes in "quasicontinuum valence subbands", i.e., in subbands at

$$\dot{E}_h < E_{F_p} - E_A - 2V_{Dn} \qquad (4.2.7)$$

(see, Fig. 4.12.a), however, may become quite large. The interband matrix elements for recombination with relaxed holes or electrons, respectively, do not contain exponentially small overlap factors. Therefore, the radiative lifetimes for these "vertical recombination" processes are of the order of typical bulk values. It is clear that the fraction of recombination processes at $\hbar\omega > E_{F_n} - E_{F_p}$ and their spectral distribution depends strongly on the excitation photon energy in the case of photoluminescence. The "vertical" processes, for example, can only be induced by two-photon or two-step excitation processes, if $\hbar\omega^{exc} < E_g - E_A$ or $< E_g + E_{c,0} + E_{F,0}$ |86|.

For several reasons the radiative recombination of electrons and holes across the indirect gap in real space is also interesting with respect to stimulated processes:

(i) Population inversion within a wide energy range can be achieved at low excitation or injection intensities. This is a consequence of the enhanced recombination lifetimes yielding unusually high steady-state electron and hole concentrations.

(ii). The absolute value of the gain or of "negative absorption" slightly below the absorption threshold $\hbar\omega = \Delta E_{F_{np}}$ will only be somewhat smaller than the corresponding absorption slightly above the threshold. Therefore, reasonable large values of optical gain at photon energies quite far below E_g may be obtained in doping SLs with high doping concentrations.

(iii) The gain is not restricted to a two-dimensional region, but it is a bulk phenomenon.

(iv) The gain spectrum, of course, is tunable.

The study of the optical properties of various SL systems yielded a lot of information on their electronic properties, which present a variety of original phenomena. However, despite the considerable volume of work devoted to optical studies of SLs, still many problems remained unresolved |44|, |54|, |90|, |91|.

4.3. Electrical transport in superlattices

Most of the interest in research on electrical transport in superlattices concerns carrier current flow parallel to the interfaces in these structures. This is a result of the enormous enhancement of carrier mobilities achieved in modulation-doped (MD) superlattices |92-95| (see, Figs. 1.7 and 1.8). Since the first announcement of MD in 1978 |96|, initial studies have been extended and a variety of investigations have been performed on these structures. Modulation-doping has been applied to single heterojunction interfaces grown by MBE, which has resulted in ultrafast field effect transistors |97|, and proved to be successful also in equivalent p-type structures with strongly improved hole mobilities |95|. Therefore, despite the fact that the so-called Bloch propagation of electrons in the direction of the superlattice axis and tunnelling processes through the wide gap potential barriers have been proposed about fifteen years ago by Esaki and Tsu in their pioneering work |98|, the propagation of charge carriers in the 2DEG or 2DHG physical conditions dominate the research works on electrical transport in superlattices.

We will present here in short the basic physical picture of the electrical transport in both, the classical (parallel to the interfaces) and quantum (perpendicular to the interfaces)modes.

4.3.1. Carrier mobilities in modulation-doped heterostructures

Since the observation of mobility enhancement in MD GaAs-AlGaAs heterojunctions, the number of publications on the subject, and researchers working in this field, has increased as much as the peak mobilities reported for these systems |99|. Starting from a mere $2\cdot10^4$ cm^2/Vs at low temperatures in 1978, mobilities now exceed 10^6 cm^2/Vs in structures with 2DEG densities equivalent to $\approx 10^{17}$ cm^{-3}.

Mobility enhancement in modulation-doped heterostructures results from a separation of mobile carriers from their ionized parent impurities (see,Sec.1.4). This is achieved at a heterojunction interface where exclusively the material with the wider band gap (normally AlGaAs) is doped in the vicinity of the junction while the neighbouring narrow-gap material (normally GaAs) is kept free from dopants. MBE is the preferred process to synthesize these materials. At the heterojunction, the carriers from the impurities transfer across the interface to the lower lying band-edge state of the neighbouring semiconductor forming a two-dimensional electron or hole system spatially separated from their ionized parent donors or acceptors, respectively. Conduction along the heterointerface is considerably improved over conduction in bulk material of equivalent carrier concentration mostly due to the reduction in ionized impurity scattering. Consequently, the MD advantage is predominantly present at low temperatures where in regularly doped semiconductors carrier scattering by ionized impurities prevails. At higher temperature where phonon scattering dominates, the MD advantage is reduced, though still present even at room temperature. This basic concept of MD is still unchanged today and the separation of carriers from ionized impurities remains the major cause of mobility enhancement in these structures. However subsequent studies, theoretical as well as experimental, were able to isolate remaining scattering processes and in some cases found methods to reduce their influence.

In order to discuss the improvements which the MD process has experienced during the past years, we will list briefly the major remaining scattering process for carriers in the 2DEG at the heterojunction interface. We will focus on the n-type GaAs-AlGaAs heterojunction since most of the studies have been performed on this system (see,Fig. 4.13).

(i) Scattering with bulk phonons is intrinsic and cannot be prevented at finite temperatures. The actual phonon scattering rate of carriers in MD structures might be slightly modified as compared to transport in bulk material but the deviation is expected to be small. Theoretical calculations predict scattering rates close to bulk scattering rates |99|. At temperatures higher than $\sim$ 80 K where phonon scattering prevails mobility data on the best MD structures coincide with mobility data on highly pure bulk GaAs. This agreement shows that even at high temperatures mobilities in MD structures with carrier concentrations equivalent to 10^{17} cm^{-3} can be equal to or possibly exceed mobilities of carriers in the best ($\sim 10^{13}$ cm^{-3}) bulk GaAs. Moreover, it suggests that the high--temperature (i.e.phonon) scattering processes are very similar in both structures.

(ii) Scattering with interface phonon is a possible scattering mechanism. However, since both materials (GaAs and AlGaAs) are very similar in density as well as in dielectric constant interface phonons will be weak and electron scattering via these phonons will be negligible.

(iii) The conduction band discontinuity at the interface is typically 300 meV. Two-dimensional carriers bound to the GaAs side of

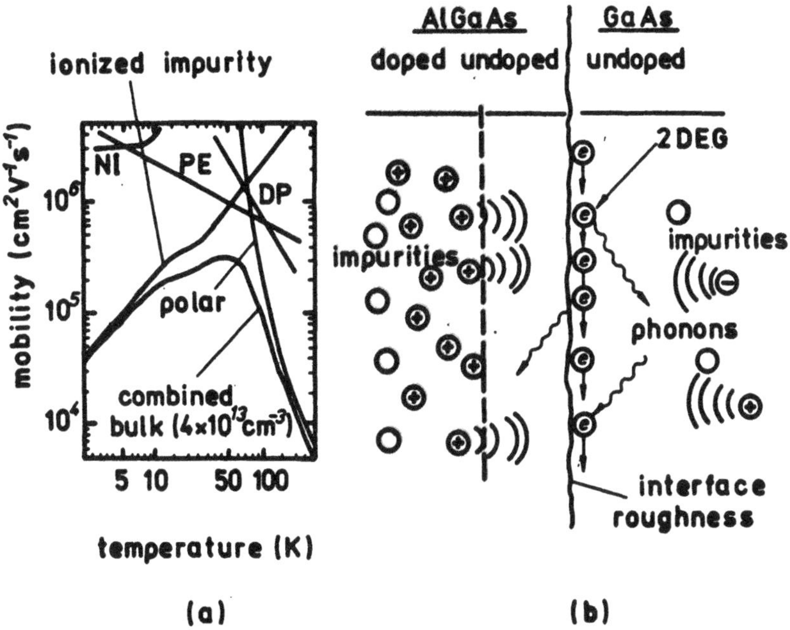

Fig.4.13. a) Scattering mechanisms in bulk GaAs for highly pure specimen (PE-piezo-electric scattering, POLAR-polar phonon scattering, DP-deformation-potential scattering, NI-neutral impurity scattering).
b) Schematic of some of the remaining scattering processes of a 2DEG at a GaAs-AlGaAs heterojunction interface An undoped AlGaAs spacer layer is included there. (after Störmer,Ref.99).

the interface therefore can tunnel into the (AlGa)As layer and are scattered by the random alloy potential of this ternary compound. However, the penetration depth is small and hence contribution to the total scattering rate will be small. The influence of the Al mole fraction (x) in (AlGa)As on the mobility in MD structures has been studied experimentally |99|. In the range from x<0.4 where (AlGa)As remains a direct semiconductor, a monotonic increase as well as monotonic decrease in mobility have been reported |99| In any case, it is unclear if the mobility dependence on Al mole fraction can decide on the importance of alloy scattering in MD structures since an increase in Al concentration reduces electron penetration but at the same time increases the potential fluctuation in the (AlGa)As. Moreover, the Al mole fraction may well govern the interface quality and influence the electron mobility via interface roughness scattering.

(iv) Scattering caused by roughness of the interface between GaAs and (AlGa)As becomes important once mobilities exceeded $\sim 10^5 cm^2/Vs$ Studies on the mobility dependence on crystal growth temperature seem to indicate that optimum mobilities are achieved at growth temperatures T_g between ~600 and ~700°C. For higher and lower T_g the mobility decays rapidly. This temperature range for optimum mobility coincides with the range of temperatures over which the surface roughness in undoped GaAs-(AlGa)As superlattices has been found to be smallest. This correlation suggests that the crystal growth temperature affects the mobility in the two-dimensional channel predominantly via variations in the degree of interface roughness.

(v) Scattering of carriers by the Coulomb tails of the remote ionized impurities in the (AlGa)As remains a major scattering process. The introduction of an undoped (AlGa)As spacer layer between the highly doped (AlGa)As and the GaAs strongly suppressed this scattering process |94|. Systematic studies of the mobility dependence on spacer thickness in single MD heterostructures seemed to reveal an unexpectedly complex behaviour. With increasing thickness of the spacer layer (d_3), mobilities increased up to $d_3 \approx 7$ nm or $d_3 \approx 10$ nm from whereon they decreased again. It seems to be understood that the decay of mobilities at larger d_3 results from a decrease in carrier density caused by a reduction in carrier transfer from the impurities to the channel for larger spacer thicknesses. Experimentally it has been found that mobilities are strongly density dependent. In general, higher densities lead to higher mobilities hence a decrease in carrier density for larger spacer thickness results in a decrease in mobility.

(vi) Impurities in the GaAs layers can be a source of carrier scattering in MD structures. Depending on the nature of the impurity and the compensation of the GaAs, these can be either charged or neutral. Charged impurities are strong scatterers while neutral impurities are weak scatterers. The GaAs layers in MD structures have generally less than 10^{15} cm^{-3} impurities [as compared to 10^{18} cm^{-3} in (AlGa)As] which are not expected to contribute considerably to the overall mobility performance of the two-dimensional electron system. Intentional doping of the GaAs layers to levels above 10^{16} cm^{-3} reduces the mobility.

Reduction of the major scattering sources for electrons in the two-dimensional channel in MD structures by careful material design has considerably improved electron mobilities in MD heterostructures. In addition to a reduction of the above listed scattering mechanisms, there is another parameter which is of paramount importance to achieve high mobility. This is the density of the two-dimensional electron system. A close correlation between electron concentration and mobility has been found experimentally. Though the actual density-dependence of the mobility varies among authors, it is generally accepted that in the range of density from $\sim 1 \cdot 10^{11}$ to $5 \cdot 10^{11}$ cm^{-2} the mobility increases strongly with density. The scattering rate for elastic scattering is strongly weighted toward large angles. This means, that the important factor is how much the electron momentum is changed along the electric field by the scattering event, not simply how frequently it is scattered. Thus for a particular fixed scattering process with some amplitude at wave vector $\vec{q}$, the scattering is relatively less effective for large k_F than for small k_F, hence the mobility increases with k_F and hence with density of the 2DEG |99|.

At present, the role of different scattering processes limiting the two-dimensional electrons mobility is well understood |95|. Considerably much less effort has been directed towards the study of two-dimensional hole systems, although it is believed that such systems are interesting not only from the point of view of fundamental physics, but also because they may have important practical applications. In a number of recently published papers, various aspects of the 2DHG in GaAs-(AlGa)As have been studied |100|. It has been shown that improved MD techniques lead to increased hole mobilities |101|. To date the highest reported hole mobility is $\sim 9.7 \cdot 10^{4}$ cm^{2}/Vs at 4.2 K |101|. Although experimental data on hole mobility are accumulating, there has been no attempt to formulate a theoretical model which could explain the existing data and predict hole mobility limits in p-type MD heterostructures. However, the present understanding of the electronic structure of p-type

GaAs-(AlGa)As MD heterostructures allows the formulation of a relatively simple model of the hole mobility |100|. The model assumes that the holes are confined in a triangular quantum well on the GaAs side of the interface. A lack of the inversion symmetry and strong spin-orbit coupling in this system results in splitting of the fourfold degenerate valence band into a series of subbands. The two uppermost heavy-hole spin subbands are degenerate for $\vec{k} = 0$ and are well separated from the lower-lying subbands so that for typical hole concentrations, only the highest subbands are occupied. Because these two subbands have different spins, there exists no substantial intersubband scattering for all the important scattering processes. Scattering rates can be calculated separately for each subband. The analysis taking into consideration all major scattering processes in the two uppermost spin subbands has shown that at large spacer widths, the predicted hole mobility limit exceeds $2 \cdot 10^5$ cm^2/Vs and is entirely determined by acoustic -phonon deformation-potential scattering |100|.

4.3.2. Resonant tunneling and hopping mobility in compositional superlattices

The current and electrical conductance as functions of the voltage applied in the direction of the type I superlattice axis show nonlinear characteristics and exhibit an oscillatory behaviour beyond a certain threshold voltage |102|. The differential conductance in the direction of the SL axis first gradually decreases, followed by a rapid drop to negative values, then at high fields, exhibits an oscillatory behaviour with respect to the applied voltages. The observed period of the oscillation in terms of applied voltage appears to coincide with the energy difference between the quantized states or bands.

The observed characteristics are interpreted as schematically illustrated in Fig. 4.14. At low applied fields, a marginal band-type conduction governs the electron transport. The mobility deduced from the conductance near zero voltage is of the order of 50 cm^2/Vs, in the direction of the superlattices axis. After showing a slow decrease as in Fig. 4.14a, the conductance starts to drop rapidly at a field of $\sim 10^4$ V/cm. The voltage drop per unite cell (per one SL period) approaches about 8 meV, if a uniform field distribution is assumed. Beyond this point it seems reasonable to assume that

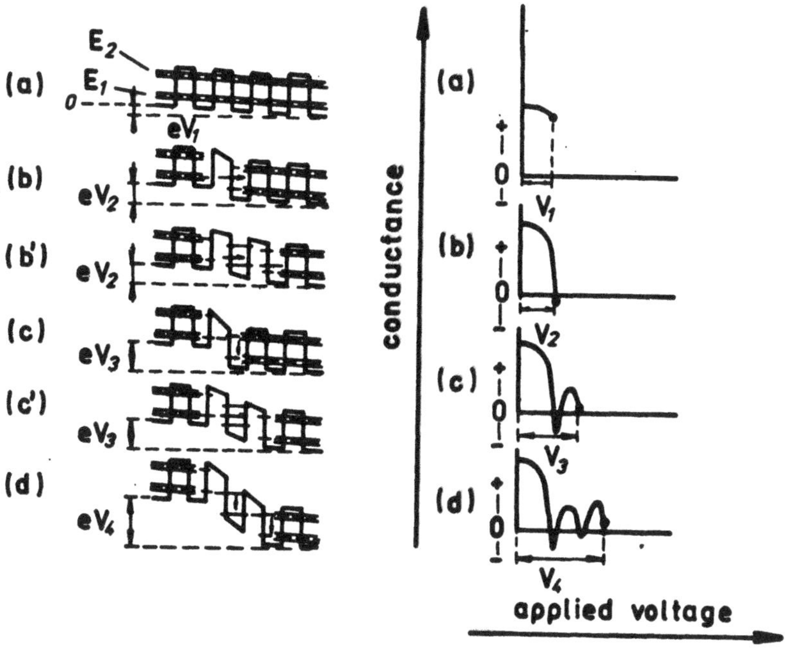

Fig.4.14. Schematic energy diagrams (left side) and corresponding conductances (right side) of a type I SL (after Esaki et al.,Ref.102)

the band conduction fails to be sustained throughout the entire SL region and a narrow high-field domain is spontaneously generated. Although the domain region may initially extend only to one or two carriers (Fig.4.14b and b'), a substantial fraction of the total voltage will be applied across it. This will leave intact the band conduction in the rest of the SL region. It should be pointed out that the high-field domain formation, initiated possibly by a random noise fluctuation or more likely an unavoidable nonuniformity in the SL structure, is indeed an inherent feature of the voltage-controlled negative-conductance medium.
We proceed to analyse tunneling characteristics across the high-field domain, because it will dominate the total conductance characteristic of the SL system. As shown in Figs.4.14b and c or in

Figs. 4.14b' and c', one may realize that this situation is somewhat analogous to the double-barrier tunneling |103|. In the most simple case, a single barrier is sufficient to provide a current peak or a negative conductance arising from matching or mismatching of energy levels on both sides of the domain, as shown in Fig.4.14c, because all involved electrons here are two-dimensional in nature. In this domain tunneling, the current peak or dip is expected to be just as sharp as that in the case of the double barrier. The tunneling current across the domain will increase at the matching condition when the applied voltage across it reaches a value corresponding to the difference between E_1 and E_2, as shown in Fig. 4.14c and c'. With further increase in applied voltage, the domain region will be expanded to the adjacent barriers, one by one, and correspondingly, off- and on-matching conditions will be alternately repeated. Fig. 4.14d shows that both barriers in the domain are simultaneously at the matching condition. This discrete nature in the domain expansion will give the oscillatory behaviour in the conductance curve, as illustrated sequentially from top to bottom on the right-band side of Fig. 4.14. Therefore, the observed voltage period (0.21-0.24 V) will provide the experimental value for E_2-E_1, which is in good egreement with the experimental data obtained from optical measurements.

The observed negative conductance is not due to the Gunn effect, because the onset voltage (0.4-0.8 V) for the negative conductance, the electron mobility (50 cm^2/Vs), and the electron mean free path (a few hundred angstroms) in the structure are much too low to activate the electron transfer mechanism. Furthermore, the observation of the oscillatory effect will exclude the possibility of the involvement of hot electrons. The involvement of the optical phonons, $\hbar\omega_o$, is also clearly denied in that case |102|.

Let us now consider the low field mobility μ_z of carriers in the direction perpendicular to the interfaces (SL axis direction). Different band theories have been used to predict the superlattice transport properties in this direction |104|. For a typical $Al_xGa_{1-x}As$ type structure, rough estimates give μ_z not drastically different from bulk values. The effective mass along z-direction is heavier but stays of the order of the alloy value if subbandwidth is higher than, say 50 meV. Such relatively large subbandwidths are theoretically possible at the lowest reasonable barriers thicknesses (∿2 nm). On the contrary, if we suppose that electrons are localized in the layers, only marginal effects give rise to a conduction process. Much lower μ_z values would be obtained as the conductivity is only governed by tunnel type hops from layer to layer. The main effect responsible for that is probably the potential disorder due to impurities or interface states which forbid the existence of the Bloch coherent state.

The experimental access to μ_z is not as easy as for μ_t since the total growth thickness of a superlattice can hardly be more than 1-2 μm. Therefore, the actual electric field chart in the sample cannot be controlled. We will present here the results obtained on three-layer structures made of two n^+-GaAs contact layers on either side of a $Ga_{1-x}Al_xAs$/GaAs superlattice |104|. The samples were prepared by MBE on a laser-quality n^+-GaAs substrate. After growth and characterization, the epilayers were processed to allow for transport measurements. The central part of the process is a AuGe ohmic contact deposit and anneal, to obtain contact resistivity less than 10^{-4} Ωcm.
A question may arise about the annealing phase needed for such ohmic contacts. It is essential that any process stage should not destroy the superlattice structure. The X-ray studies on annealed SLs tend to prove that the required annealing phase (350-450°C during a few minutes) does not result in any significant Al interdiffusion. After the contact process, samples were cut into chips. These chips were on different sample holders, according to the type of transport measurement they were designed for. Three transport experiments were considered: the I-V (current-voltage) characteristics, the high-frequency conductivity, and the time of flight optical measurement.

Two types of I-V characteristics are observed in these samples. These were, either a nearly linear law $I \sim V^n$ ($1<n<2$) or an exponential increase of I with V which cannot be fitted to any expected law in such structures. In most cases it was not possible to decide whether the second type of the I-V characteristic was due to ohmic contact poorness or to a pinning of the Fermi level in the structure. The $I \sim V^n$ characteristic can be, however, correctly interpreted by injection of electrons from a n^+ cathode into either a subband or in a system of localized states. To interpret this with no apriori idea about the actual electric field chart in the layers, use was made of a numerical simulation of the $n^+/SL/n^+$ structure in which the SL is replaced by an effective medium of adjustable macroscopic parameters. On the basis of the I-V measurements it may be concluded, that in doped superlattices coherent Bloch states are not possible. The main reason is that an impurity random potential produces a strong localization into the layers.

Another experimental determination of μ_z is the high-frequency measurement, using a particular sample holder. These measurements were made at a frequency f sufficiently high such that the non dissipative space charges at the SL/n^+ contacts were eliminated by that means. However, such measurements are less precise than the above fit of I-V characteristics. Preliminary results |104| on the same sample series have been found to be in good agreement with the I-V data.

Finally, a time of flight technique was developed. A somewhat different photolithographic process is used to obtain a photodiode geometry. Carriers injected near one SL end are swept out more or less quickly by the applied electric field. A numerical analysis technique was used. Preliminary results and interpretation on a sample made with the same epitaxial wafer are in agreement with the previous data.
The temperature was not systematically varied, as it would require to introduce trapping in the injection modelling. However, the low-field data taken on a few samples all indicate that μ_z decreases by a factor 4-5, in good agreement with the localized state prediction.

The low-field conductivity and I-V characteristics of the considered $n^+/SL/n^+$ structures were interpreted |104| with respect to two opposite theories: the Bloch propagation of electrons in the first subband, and the hopping conductivity from layer to layer. The I-V characteristics were interpreted in terms of electron injection from the n^+ cathode into the multilayer SL system, and clearly indicated that in doped SL with $N_D \sim 10^{17}$ cm^{-3}, experimental results are in agreement with a localized states mobility calculation. For undoped samples, the theoretical interpretation may be more complex as the trapping level due to interface disorder would dominate the transport properties.

Resonant tunneling behaviour for electrical transport has also been observed in type II superlattices |67|, as well as in strained-layer SLs |105|. However, the experimental data related to this problem are far from being complete. This is the reason why we will no longer dwell on this problem here.

4.3.3. Electrical transport in doping superlattices

The conductivity in doping superlattices differs in two fundamental points from their heterostructure counterparts. The transport parallel to the layers is tunable within wide limits and generally involves electrons and holes. The transport in the direction of periodicity in the case of nipi-semimetals (see,Fig.2.15) is dominated by electron-hole generation and recombination processes between n-type and p-type layers.

Let us begin with the transport parallel to the "interfaces" in doping SLs. The conductivity of a doping SL in the direction parallel to the layers contains an electron contribution

$$\sigma_n = e\,\mu_n(n^{(2)})n^{(2)}/d \tag{4.3.1}$$

and a hole contribution

$$\sigma_p = e\,\mu_p(p^{(2)})p^{(2)}/d \tag{4.3.2}$$

These electron and hole contributions to the transport may be separated from each other completely by the use of selective electrodes.

Selective n-type electrodes (in practice to be accomplished by a region of heavy n doping including all the doping layers) form ohmic contacts only to the n-type layers, whereas they provide blocking pn junctions with respect to the p-type layers. Thus, a pair of selective n-type electrodes allows the measurement of the electron contribution σ_n, and, correspondingly, a pair of selective p-type electrodes does for σ_p.

The relation between effective band gap $E_g^{eff,n}$ (or between the difference in quasi-Fermi levels $\Delta E_{F_{np}} = E_{F_n} - E_{F_p}$) and carrier concentration per layer implies that an external bias U_{np} applied between a selective n and p electrode causes the injection or extraction of electrons and holes until $n^{(2)}$ and $p^{(2)}$ correspond to a quasi-Fermi level difference

$$\Delta E_{F_{np}} = eU_{np} \qquad (4.3.3)$$

This is true as long as eU_{np} is larger than its threshold value eU_{np}^{th} at which $n^{(2)}$ or $p^{(2)}$, whichever is smaller, becomes zero. For $eU_{np}<eU_{np}^{th}$, also the non-vanishing carrier concentration does not change anymore, because of the requirement of charge neutrality.

The variation of σ_n and σ_p due to modulation of $n^{(2)}$ and $p^{(2)}$, respectively, is quite similar to the tuning of the two-dimensional channel conductivity in inversion layers as a function of gate voltage. In particular, this applies to such phenomena as the variation of mobility as a function of carrier concentration, to the influence of higher subband population on mobility and on the magnetoconductivity oscillations, which have been studied intensively for those structures |106|.

In the present case, however, both types of carriers are affected by the modulation and, although the system is dynamically two-dimensional, it is spatially three-dimensional. A thermodynamically interesting consequence is that a non-equilibrium state with $np<n_0p_0$, i.e. with $\Delta E_{F_{np}}<0$, is metastable in the whole bulk of an nipi-semimetal, corresponding to a negative effective gap. Normally, non-equilibrium situations with carrier deficiency can only occur in spatially limited regions. Perhaps the most spectacular transport property of doping SLs concerns their giant photoresponse. This is a direct consequence of the extremely long recombination lifetimes of the photoexcited charge carriers |96|.
Various mechanisms for transport in the direction of periodicity which have been investigated for the case of compositional SLs are related to unipolar (usually n-type) conduction between potential wells of the same type in appropriately doped systems. Apart from these transport peculiarities which are expected for compositional SLs and for doping SLS as well, there are also specific doping SL phenomena due to bipolar transport between n-and p-type layers, if both types of layer are populated. This possibility has been discussed very briefly only in the first theoretical investigation of doping SLs |8|.

The potential profile for an nipi-semimetal with an external field F_z, applied in the direction of periodicity, is shown schematically in Fig.4.15a. The solid lines represent the situation immediately after switching on F_z at t=0. The quasi-Fermi level distri-

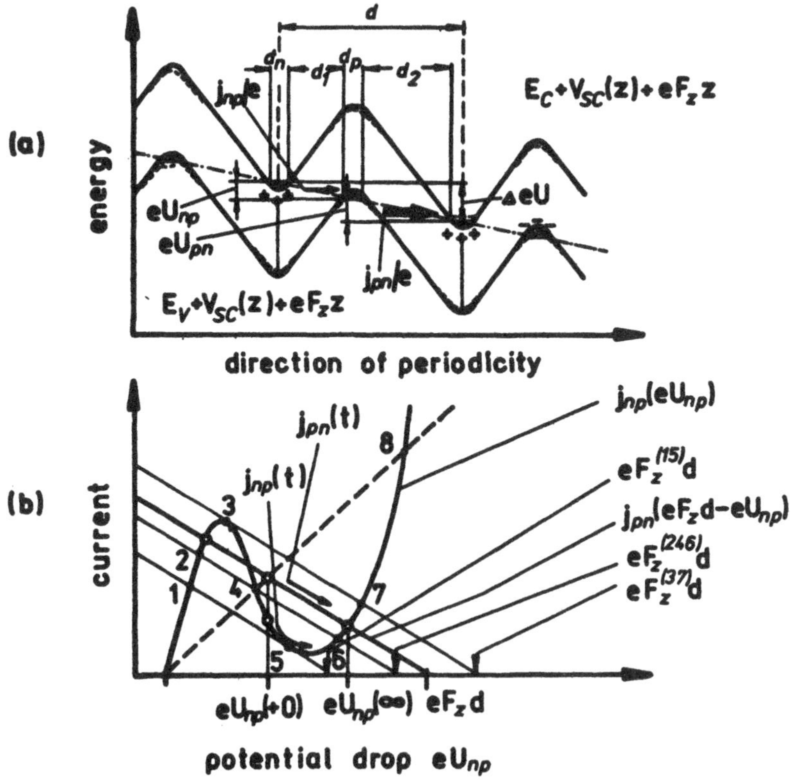

Fig.4.15. a) Semimetal-type GaAs doping SL (without inversion symmetry) with an external potential $eF_z z$ applied in the direction of periodicity.
b) Current density j_{np} versus potential drop eU_{np} between neighbouring n and p layers of a GaAs doping SL acting as a tunnel diode. The current-voltage relation for the other kind of p-n junction is approximated by a linear j-U characteristic. The relation j_{pn} versus $eF_z d-eU_{np}$ is shown for a given fixed value of F_z. The abscissa of the intersection of the dashed line with $j_{pn}(eF_z d - eU_{np})$ gives the value of $eU_{np}(+0)$ from Eq. (4.3.4) immediately after switching on the external field F_z. It fields $F_z > F_z(246)$ one has $|j_{pn}(+0)| > j_{np}(+0)|$. Therefore, according to Eq. (4.3.7) $e\dot{U}_{np}>0$, until $j_{pn}(\infty) = j_{np}(\infty)$ is reached. For $F_z<F_z(246)$, the value of eU_{np} decreases with time. At $F_z=F_z(246)$ the situation is unstable (after Ploog et al.Ref.86)

bution at t= -0 is roughly determined by the superposition of the potential eF_z to the ground-state space-charge potentials. In the case of a doping SL without inversion symmetry and with $d_n, d_p \ll d_{i,1}, d_{i,2}$, e.g. one has

$$\Delta E_{F_{np}} = eU_{np} \approx eF_z[d_{i,1} + (d_n+d_p)/2] \quad (4.3.4)$$

$$\Delta E_{F_{pn}} = eU_{pn} \approx eF_z[d_{i,2} + (d_n+d_p)/2] \quad (4.3.5)$$

and, per superlattice period

$$\Delta(eU) = eU_{np} + eU_{pn} = eF_z d \quad (4.3.6)$$

The dominant contributions to the transport are the transitions between neighbouring layers, leading to the currents $i_{np}(eU_{np})$ and $j_{pn}(eU_{pn})$. Although the system appears as a simple series of forward and reverse bias Esaki tunneling diodes |86|, it differs from such an arrangement by an additional parameter. Except for two distinct values of $\Delta(eU) = eF_z d$, the interlayer currents $j_{np}(eU_{np})$ and $j_{pn}(eU_{pn})$ with eU_{np} and eU_{pn} calculated from Eqs. (4.3.4) and (4.3.5) will differ from each other (see points 4 and 8 in Fig. 4.15b). Consequently, the space-charge density per layer and, hence, the values of eU_{np} and eU_{pn} are changing with time according to

$$\dot{n}^{(2)} = \dot{p}^{(2)} = [j_{np}(eU_{np}) - j_{pn}(eU_{pn})]/e \begin{cases} >0 \rightarrow \begin{cases} e\dot{U}_{np} < 0, \\ e\dot{U}_{pn} > 0, \end{cases} \\ <0 \rightarrow \begin{cases} e\dot{U}_{np} > 0 \\ e\dot{U}_{pn} < 0 \end{cases} \end{cases} \quad (4.3.7)$$

Thus, $eU_{np}(t)$ increases after the sudden application of an external field $F_z > F_z(246)$ (see Fig. 4.15b) until it reaches the steady-state value $eU_{np}(\infty)$ for which $j_{np}|eU_{np}(\infty)| = j_{pn}|eF_z d - eU_{np}(\infty)|$. Similarly, $eU_{np}(t)$ decreases for $F_z < F_z(246)$ until the stabile point to the left from point 3 is reached.

As yet, the theoretical and experimental work on electrical transport in doping superlattices has focused mainly on GaAs-like semiconductors having a direct energy gap at the Γ point of the Brillouin zone. New, intriguing electronic features should, however, be expected in IV-VI doping superlattices, owing to the energy band structure of the host materials which have direct energy gaps at the L point of the zone. At present, however, no experimental data are available.

4.4. Literature to chapter 4

|1| Esaki,L.L., L.L.Chang, W.E.Howard, and V.L.Rideout: Transport properties of a GaAs-AlGaAs superlattice, in: Proceedings of the XI Intern.Conf.Phys.Semiconductors, Warsaw, Poland,1972, v.1,431-437, PWN-Polish Scientific Publisher,Warsaw, and Elsevier, Amsterdam 1972

|2| Chang,L.L., L.Esaki, W.E.Howard, R.Ludeke: The growth of a GaAs-AlGaAs superlattice,J.Vac.Sci.Technol. 10 (1973) 1,11-16

|3| Davis,R.H., and H.H.Hosack:Double barrier in thin film triodes, J.Appl.Phys. 34 (1963) 4,864-866

|4| Iogansen,L.V.: On the possibility of resonance passage of electrons in crystals through a system of barriers, Zh.Eksp. Teor.Fiz., 45 (1963) 2,207-213

|5| Keldysh,L.V.: On the influence of ultrasonics on the electron spectrum of a crystal, Fiz.Tverd.Tela, 4 (1962) 8, 2265-2267

|6| Kastàlskij,A.A.: New optical and electrical effects in a standing light wave, Zh.Eksp.Teor.Fiz.Pisma Red., 10 (1969) 7,328-332

|7| Döhler,G.H.: Electron states in crystals with nipi-superstructure, Phys.Stat.Sol. (b) 52 (1972) 1,79-92

|8| Döhler,G.H.: Electrical and optical properties of crystals with nipi-structure, Phys.Stat.Sol. (b) 52 (1972) 2,533-545

|9| Ludeke,R., L.Esaki, and L.L.Chang: $Al_xGa_{1-x}As$ superlattices profiled by Auger electron spectroscopy, Appl.Phys.Lett. 24 (1974)9,417-419

|10| Segmüller,A.,P.Krishna, and L.Esaki: X-ray diffraction study of a one-dimensional GaAs-AlAs superlattice, J.Appl.Cryst. 10 (1977, 1,1-6

|11| Laidig,W.D., C.K.Peng, and Y.F.Lin: Effects of strain and layer thickness on the growth of $In_xGa_{1-x}As$-GaAs strained--layer superlattices, J.Vac.Sci.Technol. B2 (1984) 2,181-185

|12| Esaki,L., and L.L.Chang: Semiconductor superfine structures by computer-controlled MBE, Thin Solid Films 36 (1976)2, 285-298

|13| Zinner,E.: Sputter depth profiling of microelectronic structures, J.Electrochem.Soc. 130 (1983)5,199 C-222 C

14 Seah,M.P.: A review of the analysis of surfaces and thin films by AES and XPS, Vacuum 34 (1984)3/4,463-478

|15| Keehlyside,M., F.H.Stott, and G.C.Wood: Ion sputtering in the surface analysis of practical surfaces. Surface and interface Analysis 5 (1983)2,64-70

|16| Fleming,R,M., D.B.McWhan, A.C.Gossard, W.Wiegmann,and R.A. Logan:X-ray diffraction study of interdiffusion and growth in $(GaAs)_n$ $(AlAs)_m$ multilayers,J.Appl Phys 51 (1980)1 357-363

|17| Terauchi,H., S.Sekimoto, N.Sano, H.Kato, and M.Nakayama: X-ray study on impurity diffusion in a GaAs-AlAs superlattice, Appl.Phys.Lett. 44 (1984)10,971-973

|18| Halliwell,M.A.G., M.H.Lyons, B.K.Tanner, and P.Ilczyszyn: Assessment of epitaxial layers by automated scanning double axis diffractometry, J.Cryst.Growth 65 (1983)3,672-678

|19| Halliwell,M.A.G., M.H.Lyons, and M.J.Hill: The interpretation of X-ray rocking curves from III-V semiconductor device structures, J.Crystal Growth 68 (1984)2,523-531

|20| Speriosu,V.S., and T.Vreeland,Jr.: X-ray rocking curve analysis of superlattices, J.Appl.Phys. 56 (1984)6,1591-1600

|21| Speriosu,V.S., M.A.Nicolet, S.T.Picraux, and R.M.Biefeld: Depth profiles of perpendicular and parallel strain in a $GaAs_xP_{1-x}$/GaP superlattice, Appl.Phys.Lett. 45 (1984)3,223-225

|22| Quillec,M., L.Goldstein, G.LeRoux, J.Burgeat, and J.Primot: Growth conditions and characterization of InGaAs/GaAs strained-layer superlattice, J.Appl.Phys. 55 (1984)8,2904-2909

|23| Speriosu,V.S., M.A.Nicolet,J.L.Tandon, and Y.C.M.Yeh: Interfacial strain in $Al_xGa_{1-x}As$ layers on GaAs, J.Appl.Phys. 57 (1985)4,1377-1379

|24| Bartels,W.J., and W.Nijman: X-ray double-crystal diffractometry of $Ga_{1-x}Al_xAs$ epitaxial layers, J.Crystal Growth 44 (1978)5,518-525

|25| Saris,F.W., W.K.Chu, C.A.Chang, R.Ludeke, and L.Esaki: Ion backscattering and channeling study of InAs-GaSb superlattices, Appl.Phys.Lett. 37 (1980)10,931-933

|26| Chu,W.K., F.W.Saris, C.A.Chang, R.Ludeke, and L.Esaki: Ion-beam crystallography of InAs-GaSb superlattices, Phys.Rev.B 26 (1982)4,1999-2010

|27| Chu,W.K., C.K.Pan, and C.A.Chang: Superlattice interface and lattice strain measurements by ion channeling,Phys.Rev.B 28 (1983)7,4033-4036

|28| Chang,C.A., and W.K.Chu: Channeling studies of Ge-GaAs superlattices grown by MBE, Appl.Phys.Lett. 42 (1983)5,463-465

|29| Hamdi,A.H., V.S.Speriosu, M.A.Nicolet, J.L.Tandon, and Y.C.M.Yeh: Analyses of MO CVD-grown $Al_xGa_{1-x}As$/GaAs strained superlattice structures by backscattering spectrometry and X-ray rocking curves, J.Appl.Phys. 57 (1985)4,1400-1402

|30| Hamdi,A.H., V.S.Speriosu, J.L.Tandon, and M.A.Nicolet: Combined use of ion backscattering and X-ray rocking curves in the analyses of superlattices, Phys.Rev.B 31 (1985)4,2343-2347

|31| Bogh,E.: Rutherford scattering of protons in the surface layers of a tungsten single crystal, Phys.Rev.Lett. 19 (1067) 2,61-63

|32| Narusawa,T., K.L.I.Kobayashi, and H.Nakashima: Characterization of semiconductor surfaces and interfaces by high-energy ion scattering, Ext.Abstracts, 16th Intern.Conf.Solid State Devices Materials, Kobe,1984 pp.289-292

|33| Feldman,L.C.: Application of channeling to surface sciences, Physica Scripta 28 (1983)2,303-307

|34| Feldman,L.C.: MeV ion scattering for surface structure determination, Crit.Rev.Solid State Mate.Sci. 10 (1981)2,143-167

|35| Feldman,L.C., J.W.Mayer, and S.T.Picraux: Material analysis by ion channeling, Academic Press, New York 1982

|36| Chu,W.K.:Energy loss and energy straggling, in: Ion beam handbook for material analysis, J.W.Mayer, and E.Rimini (eds.), Academic Press, New York 1977

|37| Barrett,J.H.: Monte Carlo channeling calculations, Phys.Rev. B 3 (1971)5,1527-1547

|38| Gemmell,D.S.: Channeling and related effects in the motion of charged particles through crystals, Rev.Mod.Phys. 46 (1974)1,129-160

|39| Osbourn,G.C.: Strained-layer superlattices from lattice mismatched materials, J.Appl.Phys. 53 (1982)3,1586-1589

|40| Farr,M.K., J.G.Taylor, and S.K.Sinha: Lattice dynamics of GaSb, Phys.Rev.B 11 (1975)4,1587-1594

|41| Voisin,P., G.Bastard, M.Voos, E.E.Mendez, C.A.Chang, L.L. Chang, and L.Esaki: Optical transmission in GaSb-AlSb superlattices, J.Vac.Sci.Technol. B1 (1983)2,409-411

|42| Voisin,P., C.Delalande, M.Voos, L.L.Chang, A.Segmüller, C.A. Chang, and L.Esaki: Light and heavy valence subband reversal in GaSb/AlSb superlattices, Phys.Rev.B 30 (1984)4,2276-2278

|43| Pan,C.K., D.C.Zheng, T.G.Finstad, W.K.Chu, V.S.Speriosu, M.A. Nicolet, and J.H.Barrett: Structural study of GaSb/AlSb strained-layer superlattices, Phys.Rev.B 31 (1985)3,1270-1277

|44| Voisin,P.: Optical studies of semiconductor superlattices, Surface Sci. 142 (1984)2,460-473

|45| Dingle,R., A.C.Gossard, and W.Wiegmann: Direct observation of superlattice formation in a semiconductor heterostructure, Phys.Rev.Lett. 34 (1975)21,1327-1330

|46| Sai-Halasz,G.A., L.L.Chang, J.M.Welter, C.A.Chang, and L. Esaki: Optical absorption of $In_{1-x}Ga_xAs$-$GaSb_{1-y}As_y$ superlattices, Solid State Commun. 27 (1978)10,935-937

|47| Guldner,Y., J.P.Vieren, P.Voisin, M.Voos, L.L.Chang, and L. Esaki: Cyclotron resonance and far-infrared magneto-absorption experiments on semimetallic InAs-GaSb superlattices, Phys.Rev.Lett. 45 (1980)21,1719-1722

|48| Delalande,C., U.O.Ziemelis, G.Bastard, M.Voos, A.C.Gossard, and W.Wiegmann: Photoluminescence and excitation spectroscopy in coupled GaAs-AlGaAs quantum wells, Surface Sci. 142 (1984)2,498-503

|49| Petrou,A., J.Warnock, R.N.Bicknell, N.C.Giles-Taylor, and J. F.Schetzina: Photoluminescence of a $Cd_{0.55}Mn_{0.45}Te$-CdTe MQW structure in a magnetic field, Appl.Phys.Lett. 46 (1985)7, 692-694

|50| Guldner,Y., G.Bastard, J.P.Vieren, M.Voos, J.P.Faurie, and A.Million: Magneto-optics in a II-VI superlattice-HgTe/CdTe, Surface Sci. 142 (1984)2,593-597

|51| Gourley,P.L., and R.M.Biefeld: Quantum size effects in GaAs/ $GaAs_xP_{1-x}$ strained-layer superlattices, Appl.Phys.Lett. 45 (1984)7,749-751

|52| Marzin,J.Y., M.Quillec, E.V.K.Rao, G.Leroux, and L.Goldstein: Optical properties of strained $In_xGa_{1-x}As$ MQW structures, Surface Sci. 142 (1984)2,509-512

|53| Laidig,W.G., D.K.Blanks, and J.F.Schetzina: Reflectance of AlAs-GaAs and $In_{0.28}Ga_{0.72}As$-GaAs superlattices, J.Appl. Phys. 56 (1984)6,1791-1796

|54| Venkateswaran,U., M.Chandrasekhar, H.R.Chandrasekhar, T.Wolfram, R.Fischer, W.T.Masselink, and H.Morkoc: Photoluminescence studies of a GaAs-$Al_xGa_{1-x}As$ superlattice at 8-300 K under hydrostatic pressure 0-70 kbar, Phys.Rev. B 31 (1985) 6,4106-4109

|55| Manuel,P., G.A.Sai-Halasz, L.L.Chang, C.A.Chang, and L.Esaki: Resonant Raman scattering in a semiconductor superlattice, Phys.Rev.Lett. 37 (1976)25,1701-1704

|56| Worlock,J.M., A.C.Maciel, A.Petrou, C.H.Perry, R.L.Aggarwal, M.Smith, A.C.Gossard, and W.Wiegmann: Magneto-optical studies of two-dimensional electrons in MQW heterostructures, Surface Sci. 142 (1984)2,486-491

|57| Cerdeira,F., A.Pinczuk, J.C.Bean, B.Batlogg, and B.A.Wilson: A Raman scattering study of Ge_xSi_{1-x}/Si strained-layer superlattices, J.Vac.Sci.Technol. B3 (1985)2,600-602

|58| Colvard,C., T.A.Gant, M.V.Klein, R.Merlin, R.Fischer, H. Morkoc, and A.C.Gossard: Folded acoustic and quantized optic phonons in (GaAl)As superlattices, Phys.Rev. B 31 (1985)4, 2080-2091

|59| Jusserand,B., D.Paquet, and A.Regreny: Optical phonon quantum levels in GaAs/$Al_xGa_{1-x}As$ superlattices, Superlattices and Microstructures 1 (1985)1,61-66

|60| Colvard,C., R.Fischer, T.A.Gant, M.V.Klein, R.Merlin, H.Morkoc, and A.C.Gossard: Phonon freedom and confinement in GaAs-$Al_xGa_{1-x}As$ superlattices, Superlattices and Microstructures, 1 (1985)1,81-86

|61| Dingle,R., W.Wiegmann, and C.H.Henry: Quantum states of confined carriers in very thin $Al_xGa_{1-x}As$-GaAs-$Al_xGa_{1-x}As$ heterostructures, Phys.Rev.Lett. 33 (1974)14,827-830

|62| Weisbuch,C., R.C.Miller, R.Dingle, A.C.Gossard, and W.Wiegmann: Intrinsic radiative recombination from quantum states in GaAs-$Al_xGa_{1-x}As$ MQW structures, Solid State Commun. 37 (1981)3,219-222

|63| Dawson,P., G.Duggan, H.I.Ralph, K.Woodbridge, and G.W. t Hooft: Positions of the subband minima in GaAs-AlGaAs quantum well heterostructures, Superlattices and Microstructures 1 (1985)3,231-235

|64| Weisbuch,C., R.Dingle, A.C.Gossard, and W.Wiegmann: Optical characterization of interface disorder in GaAs-$Ga_{1-x}Al_xAs$ MQW structures, Solid State Commun. 38 (1981)8,709-712

|65| Deveaud,B., J.Y.Emery, A.Chomette, B.Lambert, and M.Bandet: Sinle monolayer well size fluctuations in the luminescence of GaAs/AlGaAs superlattices, Superlattices and Microstructures 1 (1985)3,205-208

|66| Chomette,A., B.Deveaud, J.Y.Emery, and A.Regreny: Enlarged wells as probes to study superlattices, Superlattices and Microstructures 1 (1985)3,201-204

|67| Chang,L.L., and L.Esaki: Electronic properties of InAs-GaSb superlattices, Surface Sci. 98 (1980)1,70-89

|68| Chang,L.L., G.A.Sai-Halasz, L.Esaki, and R.L.Aggarwal: Spatial separation of carriers in InAs-GaSb superlattices, J.Vac. Sci. Technol. 19 (1981)3,589-591

|69| Fasolino,A., and M.Altarelli: Landau levels and magneto-optics of semiconductor superlattices, Surface Sci. 142 (1984) 1,322-325

|70| Fritz,I.J., R.M.Biefeld, and G.C.Osbourn: Photocurrent spectroscopy of a strained-layer superlattice, Solid State Commun. 45 (1983)4,323-325

|71| Wu,G.Y., C.Mailhiot, and T.C.McGill: Optical properties of HgTe-CdTe superlattices, Appl.Phys.Lett. 46 (1985)1,72-74

|72| Herman M.A., and M.Pessa: $Hg_{1-x}Cd_xTe-Hg_{1-y}Cd_yTe$ $(0<x ,y<1)$ heterostructures-properties, epitaxy, and applications, J. Appl.Phys. 57 (1985)8,2671-2694

|73| Schulman,J.N., and T.C.McGill: The CdTe/HgTe superlattice--proposal for a new infrared material, Appl.Phys.Lett. 34 (1979)10,663-665

|74| Bastard,G.: Theoretical investigations of superlattice band structure in the envelope-function approximation, Phys.Rev.B 25 (1982)12,7584-7597

|75| Smith,D.L., T.C.McGill, and J.N.Schulman: Advantages of the HgTe/CdTe superlattice as an infrared detector material. Appl. Phys.Lett. 43 (1983)2,180-182

|76| Guldner,Y., G.Bastard, and M.Voos: Calculated temperature dependence of the band gap of HgTe/CdTe superlattices, J.Appl. Phys.57 (1985)4,1403-1405

|77| Schulman,J.N., and Y.C.Chang: HgTe/CdTe superlattice band gap enhancement due to interdiffusion, Appl.Phys.Lett. 46 (1985) 6,571-573

|78| Guldner,Y., G.Bastard, J.P.Vieren, M.Voos, J.P.Faurie, and A. Million: Magneto-optical investigations of a novel superlattice HgTe/CdTe, Phys.Rev.Lett. 51 (1983)10,907-910

|79| Faurie,J.P., M.Boukerche,S.Sivananthan, J.Reno, and C.Hsu: HgTe/CdTe superlattices grown by MBE, Superlattices and Microstructures 1 (1985)3,237-244

|80| Datta,S., J.K.Furdyna, and R.L.Gunshor: Diluted magnetic semiconductor superlattices and heterostructures, Superlattices and Microstructures 1 (1985)4,327-334

|81| Petroff,P.M., C.Weisbuch, R.Dingle, A.C.Gossard, and W.Wiegmann: Luminescence properties of $GaAs-Al_xGa_{1-x}As$ double heterostructures and MQW superlattices grown by MBE, Appl.Phys. Lett. 38 (1981)12,965-967

|82| Masselink,W.T., Y.L.Sun, R.Fischer, T.J.Drummond, Y.C.Chang, M.V.Klein, and H.Morkoc: Enhanced luminescence from AlGaAs/GaAs single quantum well structures through improved interfaces, J.Vac.Sci.Technol. B2 (1984)2,117-122

|83| Zhang,X.C., S.K.Chang, A.V.Nurmikko, L.A.Kolodziejski, R.L. Gunshor, and S.Datta: Interface localization of excitons in $CdTe/Cd_{1-x}Mn_xTe$ MQWs, Phys.Rev.B 31 (1985)6,4056-4059

|84| Yip,S.K., and Y.C.Chang: Theory of phonon dispersion relations in semiconductor superlattices, Phys.Rev. B 30 (1984) 12,7037-7059

|85| Hayes,W., and R.Loudon: Scattering of light by crystals, J. Wiley, New York 1978

|86| Ploog,K., and G.H.Döhler : Compositional and doping superlattices in III-V semiconductors, Adv.in Phys. 32 (1983)3, 285-359

|87| Jung,H., and K.Ploog: Influence of material design parameters on radiative recombination in GaAs doping superlattices grown by MBE, Appl.Phys. A 37 (1985)1,83-87

|88| Jung,H., H.Künzel, G.H.Döhler, and K.Ploog: Photoluminescence in GaAs doping superlattices, J.Appl.Phys. 54 (1983)12, 6965-6973

|89| Jung,H., G.H.Döhler, H.Künzel, K.Ploog, P.Ruden, and H.J. Stolz: Photoluminescence study of electron-hole recombination across the tunable effective gap in GaAs nipi-superlattices, Solid State Commun. 43 (1982)4,291-294

|90| Cooperman,G., L.Friedman, and W.L.Bloss: Corrections to enhanced optical nonlinearity of superlattices, Appl.Phys.Lett. 44 (1984)10,977-979

|91| Friedman,L., W.L.Bloss, and G.Cooperman: Enhanced optical nonlinearities of superlattices within the Kronig-Penney model incorporating inherent bulk nonlinearities, Superlattices and Microstructures 1 (1985)3,195-196

|92| Dingle,R., H.L.Störmer, A.C.Gossard, and W.Wiegmann: Electron mobilities in modulation-doped semiconductor heterojunction superlattices, Appl.Phys.Let. 33 (1978)7,665-667

|93| Störmer, H.L.: Modulation doping of semiconductor superlattices and interfaces, J.Phys.Soc.Japan 49 (1980) Suppl.A, 1013--1020

|94| Störmer,H.L., A.Pinczuk, A.C.Gossard, and W.Wiegmann: Influence of an undoped (AlGa)As spacer on mobility enhancement in GaAs-(AlGa)As superlattices, Appl.Phys.Lett. 38 (1981)9, 691-693

|95| Störmer,H.L.: Novel physics in two-dimensions with modulation-doped heterostructures, Surface Sci. 142 (1984)1,130-146

|96| Störmer,H.L., R.Dingle, A.C.Gossard, W.Wiegmann, and R.A. Logan: Electronic properties of modulation-doped GaAs-Al_xGa_{1-x}As superlattices, in: Proceedings of the 14-th Intern.Conf. Phys.Semicond. 1978, Edinburgh, B.L.H.Wilson (ed.) Conf.Ser. 43,557-560, Institute of Physics, London and Bristol 1979

|97| Morkoc,H.: The HEMT-a superfast transistor, IEEE Spectrum, 28-35, Febr.1984

|98| Esaki,L., and R.Tsu: Superlattice and negative differential conductivity in semiconductors, IBM J.Res.Develop. 14 (1970) 1,61-65

|99| Störmer,H.L.: Electron mobilities in modulation-doped GaAs (AlGa)As heterostructures, Surface Sci. 132 (1983)1/3,519-526

|100| Walukiewicz,W.: Hole mobility in modulation-doped heterostructures-GaAs/AlGaAs, Phys.Rev.B 31 (1985)8,5557-5560

|101| Wang,W.I., E.E.Mendez, and F.Stern: High mobility hole gas and valence band offset in modulation doped p-AlGaAs/GaAs heterojunctions, Appl.Phys.Lett. 45 (1984)6,639-641

|102| Esaki,L., and L.L.Chang: New transport phenomena in a semiconductor superlattice, Phys.Rev.Lett. 33 (1974)8,495-497

|103| Vassel,M.O., J.Lee, and H.F.Lockwood: Multibarrier tunneling in $Ga_{1-x}Al_xAs$/GaAs heterostructures, J.Appl.Phys. 54 (1983) 9,5206-5213

|104| Palmier,J.F., H.Le Person, C.Minot, A.Chomette, A.Regreny, and D.Calecki: Hopping mobility in semiconductor superlattices, Superlattices and Microstructures 1 (1985)1,67-72

|105| Gavrilovic,P., J.M.Brown, R.W.Kaliski, N.Holonyak Jr., K. Hess, M.J.Ludowise, W.T.Dietze, and C.R.Lewis: Resonant tunneling in a $GaAs_{1-x}P_x$ strained-layer quantum well heterostructure, Solid State Commun. 52 (1984)3,237-239

|106| Ando,T., A.B.Fowler, and F.Stern: Electronic properties of two-dimensional systems, Rev.Mod.Phys. 54 (1982)2,437-672

Chapter 5. DEVICE APPLICATIONS OF SUPERLATTICE STRUCTURES

Many interesting application possibilities of superlattice structures have been demonstrated recently in different semiconductor devices |1-3|. The operating principles of these devices can be reduced to one (or more) of the following concepts, namely |4|:
(i) the quantum confinement of carriers |5| and light |6|,
(ii) the spatial separation of electrons from holes or free charge carriers from their parent impurities |7|,
(iii) the rapid transfer of carriers across the boundary between thin layers |8|,
(iv) the tunneling and associated resonance of quantum mechanical wave functions |9|.
The variety of semiconductor devices with superlattices may be divided into three main groups. The first group is created by optoelectronic devices, namely, photodetectors, light emitting devices, and optical passive devices. The second group is created by the negative resistance devices. The different kinds of transistors with superlattices belong to the third group.

A detailed discussion of all kinds of semiconductor devices with superlattices exceeds far beyond the scope of this book. Therefore, in this chapter only the most important device applications of superlattice structures will be presented.

5.1. Optoelectronic devices with superlattices and multi-quantum well structures

The characteristic feature of optoelectronics is that, instead of the electronic channel used in conventional electronics, it makes use of an optical channel for transmission and processing of data. The basic semiconductor elements of an optical data transmission channel are: an electrically controlled source of electromagnetic radiation, (i.e., light emitting diode or junction laser) an element that processes or transmits optical signals (e.g., waveguide, modulator, deflector, directional coupler, etc.) and an element that converts light into electric signals (namely, a photodetector). So far, superlattices have been applied with success in light sources and in photodetectors. Therefore, the application of SL structures in these two groups of devices will be discussed in more details.

5.1.1. Injection lasers and light emitting diodes

One attractive feature of a superlattice is that its density-of-states function ρ(E), which is the most important parameter for determining the electronic and optical properties of semiconductor crystals, can be tailored by controlling the atomic arrangement of the constituent elements. For any bulk crystal or alloy, ρ(E) is always parabolic with ρ(E) = 0 at the band-edge. On the other hand, in a superlattice structure, ρ(E) is step-like with a finite value even at minimum energy. This step-like density-of-states makes the superlattice structure superior in optical properties to bulk crystals |10|. Taking advantage of this, a very low threshold laser diode has already been demonstrated in which a multi-quantum-well structure has been adopted in the active region of a double heterostructure (DH) laser diode |11|.

The MQW laser has in its active region a superlattice segment which is anisotropic in structure. This fact leads to the following advantageous device properties of MQW lasers |10|.

Consideration of both the selection rule in the electron-to-hole recombination in this anisotropic structure, and of the energy separation between heavy and light holes in the MQW structure, lead to the expectation that the optical gain for a TE (transverse electric) polarized wave in the MQW waveguide would be much larger than that for a TM (transverse magnetic) polarized wave. The experimental results of Ref. 10 demonstrate that the gain for the TE wave is larger by as much as 120 cm^{-1} at the threshold than in the case for the TM wave, whereas the difference in a conventional DH structure is 20 cm^{-1} at most. This large anisotropy in optical gain for the MQW structure makes the MQW laser attractive as a polarization-stable optical source.

An MQW passive waveguide exhibits a rather steep optical absorption spectrum compared with the conventional DH waveguide case. This is a function of the step-like density-of-states of the two-dimensional electrons and holes. Therefore, the optical absorption at lasing wavelength is much smaller in the MQW structure than in the conventional DH structure. This means that an MQW waveguide is suitable for all optical integrated circuits, including both active and passive waveguide elements on a wafer. Due to the step-like density-of-states, the optical gain bandwidth for the MQW structure is narrow: 1/2-1/3 that of the conventional DH structure. Thus, the MQW laser is suitable for single-longitudinal mode oscillation. This was confirmed by comparing spectral spreading under high-speed (500 Mbit/s) modulation of both an MQW and a conventional laser.

Impurity diffusion is known to cause disorder in superlattice structures and to result in changing them to a random alloy phase. This is an attractive characteristic for device fabrication. A window-stripe laser was accordingly fabricated using the Zn diffusion process. In this laser the active layers of the MQW structure have been changed to a random alloy only at both ends of the stripe. The alloyed regions have a wider band gap than does the remaining MQW region, and they protect the cleaved laser mirror from optical damage. Ultimate output power at the catastrophic damage limit increases by a factor of three, and reaches as high as 120 mW per facet.

MQW lasers have many advantages, as has been described above. They are thus attractive for use as light sources in optical communication channels. However, to prepare an MQW laser in the long-wavelength region, material systems other than GaAs-AlAs have to be developed. Candidates are the $In_xGa_{1-x}As$-$In_xAl_{1-x}As$ and GaSb-$Al_xGa_{1-x}Sb$ systems. In the former case, a potential well for the MQW structure is made from a ternary alloy, $In_xGa_{1-x}As$, where In and Ga are distributed randomly in the group III atomic sites. This randomness in the potential well layer, only about 30 atomic layers thick, might induce broadening of quantum energy levels for electrons and holes and cancel the advantages inherent in the MQW or superlattice structures. In the GaSb-$Al_xGa_{1-x}Sb$ system, on the other hand, the potential well layer is made from a binary compound (GaSb) where there is no statistical ambiguity in the distribution of constituent atoms. Therefore, this material system is expected to have the same advantages as the GaAs-$Al_xGa_{1-x}As$ system. Superlattices of this material system are relatively easy to grow by MBE, and some of their fundamental optical properties have already been reported |10|. Recently, success was achieved in room-temperature pulse operation of a GaSb-$Al_xGa_{1-x}Sb$ current--injected MQW laser grown by MBE. The threshold current density was 8.9 kA/cm^2, and the characteristic temperature was 60 K. A fine-scale epitaxial technology such as MBE will need to be combined in the near future with some of submicron or nanometer process, such as focused ion beam technology. As a consequence, new concepts regarding materials and devices will appear. Examples are two- or three-dimensional superlattice structures and MQW distributed feedback buried heterostructure lasers |10|.

Apart from the fact that improvements can be gained in heterojunction lasers operation by resorting to a MQW or SL configuration |12-14| various fundamental effects also can be observed in these lasing structures. Some of these, including the electron-hole asymmetry and the peculiarities of carrier collection, phonon generation, alloy clustering, layer disorder, high pressure effects, and size-induced direct-indirect transitions, are exhaustively described and discussed in Ref. 15.

Superlattices may also be applied in light emitting diodes (LEDs). For example, the incorporation of a strained layer superlattice active region into such a diode eliminates the need to design the devices in a lattice matched system or resorting to compositional grading to avoid misfit disclocations. This advantage has been demonstrated in a $GaAs/GaAs_{1-x}Sb_x$ superlattice LED |16|.

Another LED with a GaAs saw-tooth superlattice have been reported in Ref.17. A schematic diagram of the edge-emitting saw-tooth superlattice LED is shown in Fig. 5.1a. The structure is grown by MBE on heavily doped n^+-GaAs substrate. The epitaxially grown layers are n- and p-type $Al_xGa_{1-x}As$, for optical confinement of the saw-tooth superlattice region. The top p^+-GaAs layer facilitates ohmic contact formation. The doping profile of the active region consisting of Dirac-delta functions is illustrated in Fig. 5.1b. This doping profile is generated by interrupting the usual "crystal growth mode" during MBE and continuing growth by the "impurity growth mode" |17|, i.e., by closing the Ga shutter and opening the respective Be- and Si-dopant shutter, while the As shutter is kept open continuously at all times. The donor and acceptor atoms are thus localised within one single atomic monolayer of the zincblende lattice. The active region consists of 20 doping spikes of two-dimensional concentration $N_D^{(2)} = N_A^{(2)} = 5\cdot 10^{12}$ cm^{-2}, separated by 10 nm undoped GaAs. The LEDs have an area of approximately 0.4 x 0.5 mm^2.

Electrons from their parent donors recombine with holes from acceptors and consequently the active region of the LED is completely

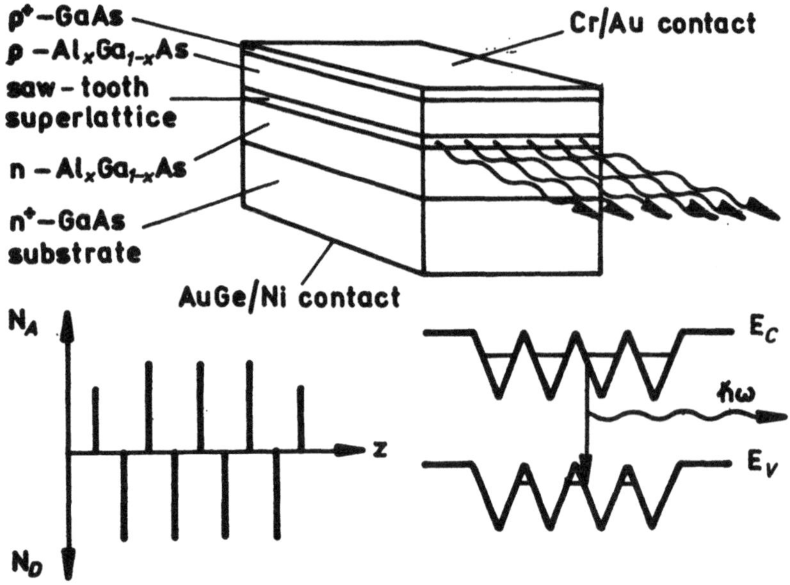

Fig.5.1. Structure of a GaAs saw-tooth superlattice light emitting diode. a) Sequence of layers grown on heavily doped n^+-GaAs substrate.
b) Doping profile of the active region, consisting of Dirac-delta functions (spike-doping)
c) Saw-tooth-shaped conduction and valence band potential in the direction of crystal growth
(after Schubert et al., Ref.17)

depleted of free carriers. The donor and acceptor charge distribution of the active region causes a saw-tooth-shaped conduction and valence band potential, as illustrated in Fig. 5.1c. Nonequilibrium electrons and holes generated by optical excitation or electrical injection occupy quantised energy levels in the V-shaped potential wells, owing to the quantum size effect. These excess electrons and holes recombine in a "quasi-vertical" mode, owing to the spatial proximity of quantised electron and hole states, resulting in a short carrier lifetime and high radiative efficiency. The superlattice bandgap energy is stable even at high excitation intensities.

Electroluminescent spectra of the LEDs operating at 300 K for two representative diodes exhibited the peak wavelengths of λ=925 nm and 965 nm, in good agreement with the values expected from the SL design parameters. These wavelengths have been inaccessible to GaAs emitting efficiency at 300 K. At the wavelength corresponding to the bandgap energy of the GaAs host material, no luminescence signal is detected. This result demonstrates the SL character of the active region of the LED. The full-width at half-maximum (FWHM) value of the electroluminescence spectrum is 60 meV. This broad experimental line-width is most likely to be due to size fluctuations of the V-shaped potential wells. The luminescent peak energy does not depend on the injection current over more than three orders of magnitude (2 mA<I<2A). This stability of the SL bandgap energy is very useful for application of the saw-tooth SL in photonic devices. The integrated electroluminescence intensity, as measured by a Si photodiode, depends linearly on the injection current, as expected for diodes emitting monochromatic but incoherent spontaneous radiation.

5.1.2. Avalanche photodiodes

A large ratio of the ionization coefficients (α,β) for electrons and holes is required for achieving low-noise at high-gain in avalanche photodiodes (APDs)|18|. Present 0.8 µm wavelength fibre-optic systems use low-noise silicon APDs, for which $\alpha/\beta \approx 50$. Future systems will operate at longer wavelengths (typically 1.3 µm) to take advantage of lower fibre losses. III-V APDs are under development for these systems. Unfortunately, α/β for most III-V compounds is approximately unity, resulting in increased avalanche multiplication noise. Thus, methods for increasing α/β in these materials are of considerable practical importance.

Two structures which enhance α/β have been demonstrated: the AlGaAs-GaAs superlattice |19| and graded gap |20| APDs. Both increase α/β in the bulk material to $\alpha \approx 5\beta$ or even to 10β.

In the SL structure, enhancement is due to band-edge discontinuity assisted impact ionization (the conduction-band discontinuities provide 25% of the electron ionization energy). To understand the SL APD consider the energy band diagram shown in Fig. 5.2a. Because of the very low doped material, the field is constant across the 2.5 µm long depletion layer. This consists of 50 alternating GaAs (45 nm) and AlGaAs (55 nm) layers. For illustrative purposes, as-

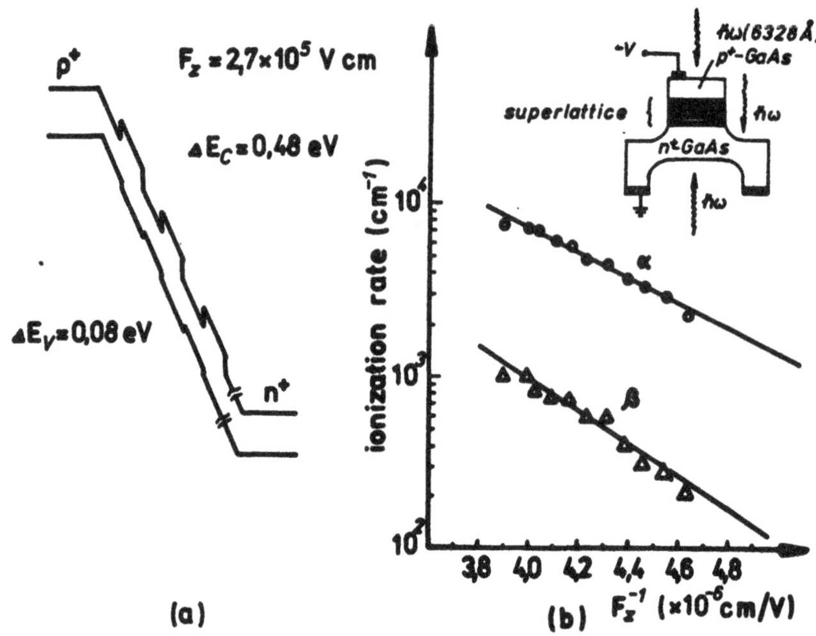

Fig.5.2. a) Energy band diagram of the superlattice APD.
b) Effective ionization coefficients for electrons ($\bar{\alpha}$) and holes ($\bar{\beta}$) in the superlattice APD. The solid lines represent least-square fits to the data. The inset shows a schematic of the APD device structure with the superlattice (after Capasso et al.,Ref.19)

sume a field $F_z = 2.7\cdot10^5$ V/cm. At this value a sizable multiplication was observed. For $F_z > 10^5$ V/cm, electrons gain between collisions an energy greater than the average energy lost per phonon scattering event (21 meV) |19|. Thus carriers are strongly heated by the field and can gain the ionization energy.

Consider now a hot-electron accelerating in an AlGaAs barrier layer. When it enters the GaAs well it abruptly gains an energy equal to the conduction band-edge discontinuity $\Delta E_C = 0.48$ eV. The effect is that the electron "sees" a ionization energy ($E_{th} \approx 1.5$eV) being reduced by ΔE_C with respect to the threshold energy in bulk GaAs ($E_{th} = 2.0$ eV). Since the impact ionization rate α increases exponentially with decreasing E_{th}, a large increase in the effective α with respect to bulk GaAs is expected. When the electron enters the next AlGaAs barrier region, the threshold energy in this material is increased by ΔE_C thus decreasing α in the AlGaAs layer. However, since $\alpha_{GaAs} >> \alpha_{AlGaAs}$, the exponential dependence on the threshold energy ensures that the average $\bar{\alpha}$ which is equal to

$$\bar{\alpha} = (\alpha_{GaAs}L_{GaAs}+\alpha_{AlGaAs}L_{AlGaAs})/(L_{GaAs}+I_{AlGaAs}) \qquad (5.1.1)$$

is largely increased (L denotes layer thicknesses).

Electrons that have impact ionized in the GaAs easily get out of the well, the voltage drop across the well is >1 V. In addition at fields > 10^5 V/cm in GaAs the average electron energy is > 0.6 eV, so that electron trapping effects in the wells are negligible.

In contrast, the hole ionization rate β is not substantially increased because the reduction in the hole ionization energy is only the valence band discontinuity of 0.08 eV. The net result is a large enhancement in the α/β ratio.

The superlattice detector wafer shown in Fig. 5.2.b (the inset) was grown by MBE |19|.

In the graded-bandgap multilayer structure, the ionization energy is provided entirely by heterointerface conduction band steps. Thus, ideally, only electrons ionize. Fig. 5.3a shows the band diagram of the graded-gap multilayer material (assumed intrinsic) at zero applied field. Each stage is linearly graded in composition from a low (E_{g1}) to a high (E_{g2}) bandgap, with an abrupt step back to low-bandgap material. The conduction band discontinuity shown accounts for most of the bandgap difference, as is typical of many III-V heterojunctions. The materials are chosen for a conduction band discontinuity comparable to, or greater than, the electron ionization energy E_{ie} in the low-gap material following the step. The band structure of the complete detector under bias is shown in Fig. 5.3b. Consider now a photoelectron generated next to the p^+ contact. Under the combination of the bias field F_z and the grading field $\Delta E_C/l$, it drifts toward the first conduction band step. The combined field $F_z-\Delta E_C/l$ is small enough so that the electron does not impact-ionize before it reaches the step. After the step, since $\Delta E_C \approx E_{ie}$, the electron impact-ionizes.

This process is repeated in each stage after each step, the elec-

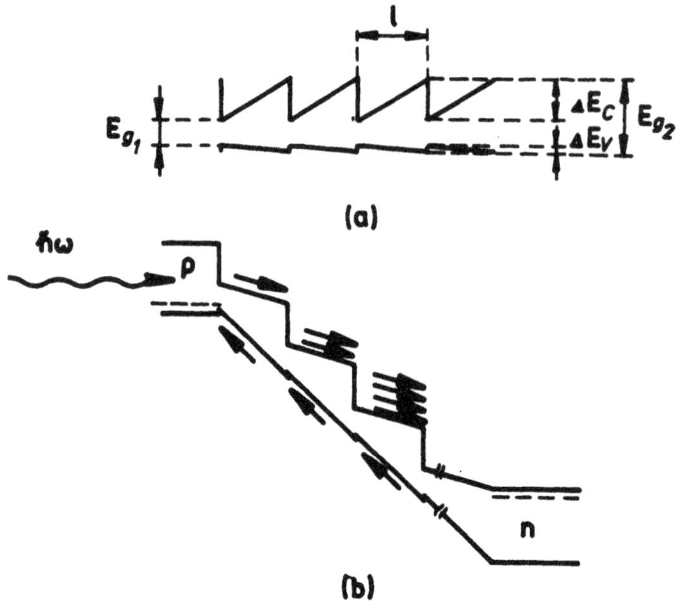

Fig.5.3. Band diagrams of the unbiased graded multilayer region (a), and the complete detector structure under bias (b), (after Williams et al.,Ref.20).

trons transverse a graded region in which they acquire the "potential energy" used for impact ionization in the low-bandgap region following the next interface. Ideally, the avalanche gain per stage is exactly 2. Each electron impact ionizes once after each conduction band step. In fact, the gain is $2-\delta$, where δ is the fraction of electrons which do not impact-ionize. The total gain of the structure is then equal to

$$M = (2-\delta)^N \tag{5.1.2}$$

where N is the number of stages. Most of the electrons impact-ionize if their energy is a tenth of an eV or so above the ionization energy E_{ie}. The competing loss mechanism is optical phonon emission. The high ionization probability is caused by the quadratic increase of the ionization cross-section above E_{ie} as a result of the density of the states and because several phonons must be emitted to reduce the electron energy below E_{ie}.

Any hole ionization is caused by the applied electric field F_z only. The valence band steps are of the wrong sign to assist ionization. For electron transport across the graded region, this bias field F_z must cancel the $\Delta E_C/l$ conduction band quasielectric field and provide a small extra component to assure drift, rather than diffusion transport. The device is then designed so that the hole ionization rate at $F_z > \Delta E_C/l$ is small. The maximum multiplication is determined by the number of such stages, which is limited by the depletion width at the lowest achievable intrinsic-region doping. Because only electrons cause ionization, this device mimics a photomultiplier. The conduction band steps correspond to dynodes.

5.1.3. A photoconductive detector based on doping superlattices

A new GaAs photodetector with high sensitivity in the whole 0.8-1.4 μm wavelength range, fabricated from totally depleted GaAs doping superlattices has been reported in Ref.21. Photoexcited electrons and holes are separated in real space by the space-charge field of the doping superlattice immediately after excitation, yielding a high quantum efficiency of this device. Because of the complete depletion, the doping superlattice behaves like a highly resistive material, which allows application of high electric field along the layers via selective n^+- and p^+-electrodes. The sensitivity of this device at 1.3 μm reaches more than 90% of the original band-edge response at 0.85 μm, and the external quantum efficiency amounts to 65% at 0.85 μm. This excellent photoresponse at longer wavelengths arises from an extremely high electric field composed of the intrinsic space charge field and applied external field, and from the existence of pronounced tail states in the forbidden gap region of the SL.

The SL configuration for the photodetector consists of 20 to 100 thin alternate n- and p-doped GaAs layers and is provided with n^+- and p^+-regions on two edges extending perpendicular to the layers to form the selective electrodes Fig. 5.4a. Through these electrodes a reverse bias voltage V_R is applied to all constituent p-n junctions of the SL. Operation of the detector can be understood by inspection of Fig. 5.4. The periodic modulation of the energy bands by the positive and negative space charge in the respective

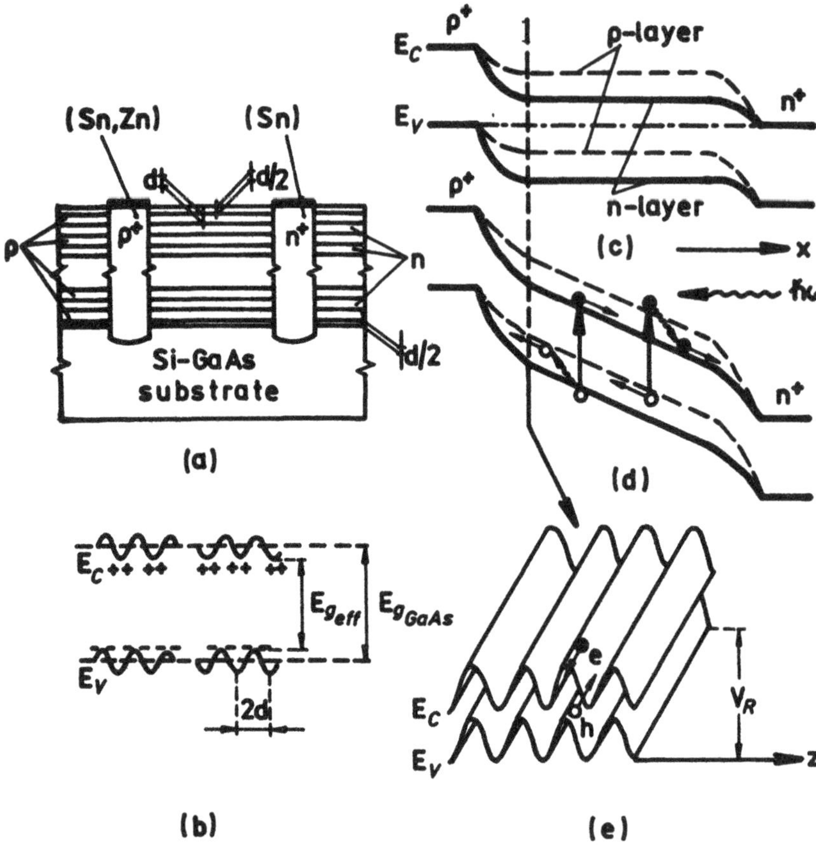

Fig.5.4. a) Layer sequence of GaAs doping Sl and arrangement of selective n^+- and p^+- electrodes.
b) periodic modulation of real-space energy bands by the positive and negative space charge in the respective layers.
c) Schematic real-space energy band diagram of a semi-insulating GaAs doping SL at zero bias in X-direction along the layers viewing the centre of each layer type.
d) The SL under operating conditions at high reverse bias with the longitudinal external field tilting the whole structure.
e) Vertical section showing the periodic modulation of the energy bands in z-direction and indicating the sweep--out of electrons and holes along the parabolic well channels in the tilted structure
(after Horikoshi et al.,Ref.21)

layers is indicated in Fig. 5.4b. A SL configuration has been chosen where the constituent layers are already totally depleted at zero bias. This requires design parameters providing equal doping densities, i.e. $N_D d_n = N_A d_p$, and an effective gap $0 < E_g^{eff} < E_g$. In this structure the intrinsic space-charge field $F_0(x)$ given by the design parameters remains constant upon variation of the reverse bias, because the donors and acceptors within the respective layers are already completely ionized at zero bias. Under these conditions the conduction and valence band-edges are flat along the layers, i.e., in the x-direction, as shown in Fig.5.4c, except for the area close to the n^+- and p^+-contact regions. Because of the total depletion of the layers, the SL is highly resistive and behaves like a semi-insulator. Therefore, when a reverse bias is applied via the selective electrodes, a constant longitudinal electric field F_1 is added in x-direction parallel to the length of the layers, i.e., the structure is tilted in layer direction, as indicated in Fig.5.4d and e. Electron-hole pairs generated by the absorption of light are effectively separated in z-direction by the strong space-charge field of the SL and then immediately swept away by the longitudinal field in x-direction to the respective electrodes. As a result, recombination of photogenerated carriers is neglegibly small and extremely high efficiencies for the absorption process are achieved. Since the photogenerated carriers have relaxation times as short as 10^{-12}s for thermalization in the conduction and valence subband systems the device speed is mainly determined by the time required to sweep out the confined electrons and holes along the parabolic well channels. In addition, due to the rapid sweep-out of the photogenerated carriers by the longitudinal electric field at high reverse bias, the bare space-charge potential of the superlattice experiences only a minor compensation from these excess carriers, and the responsivity of the device for long-wavelength irradiations does not deteriorate during operation.

5.2. Non-optoelectronic applications of superlattice structures

Among the variety of the non-optoelectronic applications of superlattice structures |2|,|22-24|, the most interesting seems to be this one, which is related to the improvement of the heterointer-

faces in III-V compound layered structures. We will discuss here only one example |25-29| but this should give the Reader sufficient information about what can be reached when SL structures are used for this purpose.
For many types of heterostructure devices the molecular beam epitaxially grown AlGaAs/GaAs system has been used to obtain impressive results. The potential of this system is considerable due to the fact that comparatively high quality heterointerfaces can be prepared using it. Its potential, however, could be further enhanced if there were a means whereby the heterointerfaces could be improved even more. From studies of MD structures, it was discovered that a substantial asymmetry exists between the quality of the inverted (binary on top of ternary) and the normal (ternary on top of binary) heterointerfaces. The quality of the normal heterointerface is usually high. In contrast, however, the inverted interface is of low quality unless some method is used to enhance its properties. Since many heterojunction devices are critically and unavoidably dependent on the inverted heterointerface, its problems can lead to seriously degraded device performance.

It has been shown |25| that significant improvement in the intrinsic photoluminescence of thin single quantum well structures can be obtained through the use of graded bulk and graded SL interfaces, and through the use of dimeric arsenic in the MBE growth process. Since the photoluminescent properties of these structures are dominated by the interfacial quality, and particularly, the inverted heterointerface, it is determined simultaneously which techniques of those studied in Ref.25 are most effective in interfacial improvement.

The structures investigated were grown on (100) oriented Si doped GaAs substrates by MBE. They are shown schematically in Fig.5.5. Three different types of interfaces were used. The first, an abrupt transition between the $Al_{0.25}Ga_{0.75}As$ confining layers and the GaAs quantum well was made. In the second, the AlAs mole fraction was intentionally graded over a thickness of ∿ 15 nm. In the third a superlattice with 2 nm layers of GaAs and 4,5, 3 and 1,5nm of $Al_{0.25}Ga_{0.75}As$ was used in such a way that the average AlAs mole fraction was decreased toward the quantum well. The thickness of the GaAs quantum well was varied in an attempt to keep the quantized levels nearly the same in all structures. The photolumines-

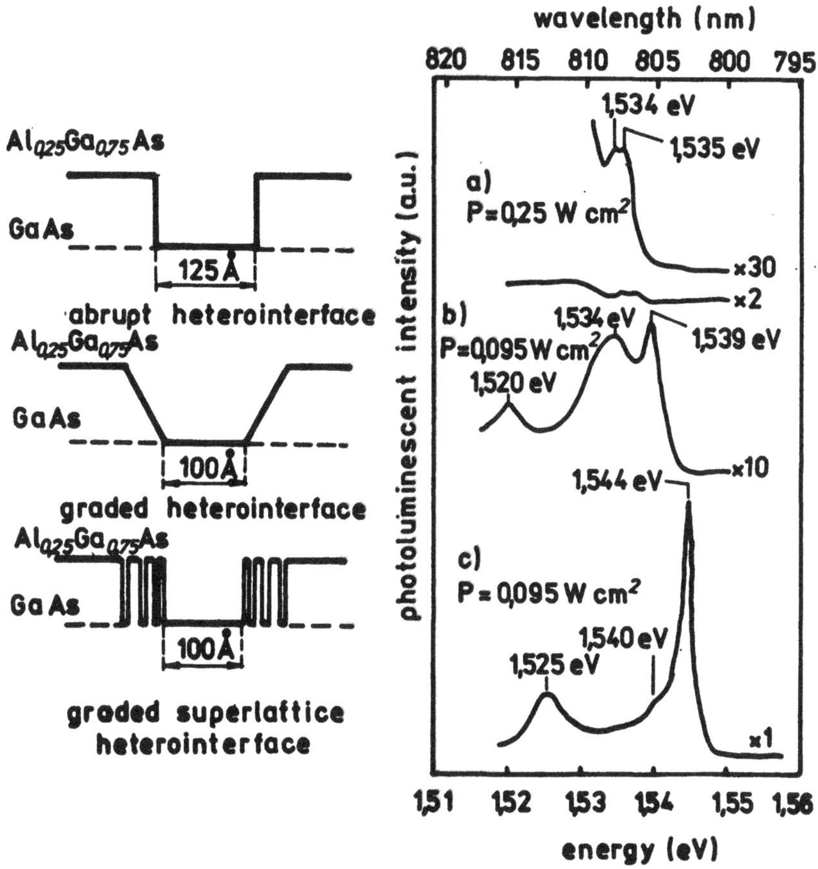

Fig.5.5. Band-edge diagrams of the quantum well structures investigated. Band bending is neglected for clarity (left side). Photoluminescence spectra from samples with a) abrupt, b) graded bulk, and c) graded superlattice interfaces. The reference is taken on the sample with graded SL interfaces (right side) (after Fischer et al.,Ref.25)

cence measurements were made with a HeNe laser (λ = 632,8 nm) in the temperature range of 2 to 70 K.

The photoluminescence spectra from the samples with the abrupt graded alloy and graded superlattice interfaces are shown in Figs.5.5a-c, respectively. In Fig. 5a, a very broad structure above 810nm is dominant, and a very weak intrinsic luminescence from the quantum well is obtained. When magnified 30 times, two peaks can be seen, one at a.534 eV and one at 1.535 eV. As will be discussed later, the 1.535 eV peak is attributed to the n = 1 heavy hole--free exciton recombination, while the one at 1.534 eV is attributed to some defect related process.

In Fig.5b, the photoluminescence spectrum from the sample with graded alloy interfaces is shown. The luminescence obtained from this sample is more intense by a factor of ∿ 8 than that in the abrupt interface sample. Also, the n = 1 heavy hole-free exciton recombination peak can clearly be seen at 1.539 eV. The peak at 1.534 eV is attributed to excitons bound to impurities.

The photoluminescence spectrum from the sample with the graded superlattice interfaces is shown in Fig. 2c. The peak response of this sample is by far the most intense by a factor of ∿ 160 over sample (a) and a factor of ∿20 over sample (c). These results clearly demonstrate that the interfacial quality plays a paramount role in determining the overall luminescent properties of the single quantum well structures. Since the luminescent properties are directly dependent on, for example, the interfacial recombination velocity (assuming such occurs by nonradiative processes), these results have immediate implications in heterostructure devices.

Literature to chapter 5

|1| Capasso,F.: New multilayer and graded gap optoelectronic and high speed devices by band gap engineering, Surface Sci. 142 (1984)1,513-528

|2| Nakagawa,T., N.J.Kawai, and K.Ohta: Design principles for CHIRP superlattice devices, Superlattices and Microstructures, 1 (1985)2,187-192

|3| Döhler,G.H.: The potential of nipi doping superlattices for novel semiconductor devices, Superlattices and Microstructures, 1 (1985)3,279-287

|4| Sakaki,H.: Evolution of new device concepts in quantum well and superlattice structures, Extended Abstracts of the 15th Conference on Solid State Devices and Materials, Tokyo,1983, pp.3-6

|5| Pallanen,M.A., D.C.Tsui, B.J.Lin, and A.C.Gossard: Localization of 2D electrons in GaAs-$Al_xGa_{1-x}As$ heterostructures, Surfaces Sci 142 (1984)1,2936

|6| Tien,P.K. Integrated optics and new wave phenomena in optical waveguides, Rev.Mod.Phys.49 (1987)2,361-420

|7| Störmer,H.L.: Novel physics in two dimensions with modulation-doped heterostructures, Surface Sci. 142 (1984)1, 130-146

|8| Capasso,F.: New device applications of band-edge discontinuities in multilayer heterojunction structures, Surface Sci. 132 (1983)2,527-539

|9| Ricco,B., and M.Ya.Azbel: Physics of resonant tunneling. The one-dimensional double-barrier case, Phys.Rev.B 29 (1984)4, 1970-1981

|10| Okamoto,H.: Optical properties of superlattice and MQW laser diode, J.Vac.Sci.Technol. B3 (1985)2,687-693

|11| Tsang,W.T.: Extremely low threshold AlGaAs modified MQW heterostructure lasers grown by MBE, Appl.Phys.Lett. 39 (1981) 11,786-788

|12| Fischer,R., J.Klem, T.J.Drumond, W.Kopp, H.Morkoc, E.Anderson, and M.Pion: Improved $Al_xGa_{1-x}As$ bulk lasers with superlattice interfaces, Appl.Phys.Lett. 44 (1984)1,1-3

|13| Holonyak,Jr,N., W.D.Laidig, M.D.Camras, J.J.Coleman, and P.D. Dapkus: Infrared-red GaAs/AlAs superlattice laser monolithically integrated in a yellow-gap cavity, Appl.Phys.Lett. 39 (1981)1,102-104

|14| Laidig,W.D., Y.F.Lin, and P.J.Caldwell: Properties of $In_xGa_{1-x}As$ strained-layer quantum-well-heterostructure injection lasers, J.Appl.Phys. 57 (1985)1,33-39

|15| Holonyak,Jr,N.: Quantum-well and superlattice lasers-fundamental effects, in: The physics of submicron structures, pp. 1-18, H.L.Grubin, K.Hess, G.J.Iafrate, D.K.Ferry (eds.), Plenum Press, New York and London 1984

|16| Klem,J., R.Fischer, W.T.Masselink, W.Kopp, and H.Morkoc: $GaAs/GaAs_{1-y}Sb_y$ superlattice light emitting diodes, J.Appl. Phys. 55 (1984)10,3843-3845

|17| Schubert,E.F., A.Fischer, and K.Ploog: GaAs saw-tooth superlattice light-emitting diode operating monochromatically at $\lambda \geqslant 0.9$ µm, Electronics Lett. 21 (1985)9,411-412

|18| McIntyre,R.J.: Multiplication noise in uniform avalanche diodes, IEEE Trans.Electron.Dev. ED-13 (1966)1,164-168

|19| Capasso,F., W.J.Tsang, A.L.Hutchinson, and G.F.Williams: Enhancement of electron impact ionization in a superlattice-a new avalanche photodiode with a large ionization rates ratio, Appl.Phys.Lett. 40 (1982)1,38-40

|20| Williams,G.F., F.Capasso, and W.T.Tsang: The graded bandgap multilayer avalanche photodiode-a new low-noise detector, IEEE Electron.Dev.Lett. EDL-3 (1982)3,71-73

|21| Horikoshi,Y., and K.Ploog: A new doping superlattice photodetector, Appl.Phys. A37 (1985)1,47-56

|22| Arnold,D., J.Klem, T.Henderson, H.Morkoc, and L.P.Erickson: Backgating in GaAs/AlGaAs modulation-doped FETs and its reduction with superlattices, Appl.Phys.Lett. 45 (1984)7, 764-766

|23| Schaff,W.J., L.F.Eastman, B.Van Rees, and B.Liles: Superlattice buffers for GaAs power MESFETs grown by MBE, J.Vac.Sci. Technol. B2 (1984)2,265-268

|24| Tischler,M.A., T.Katsuyama,N.A.El-Masry, and S.M.Bedair; Defect reduction in epitaxial layers using a GaAsP-InGaAs strained layer superlattice, Appl.Phys.Lett. 46 (1985)3, 294-296

|25| Fischer,R., W.T.Masselink, Y.L.Sun, T.J.Drummond, Y.C.Chang, M.V.Klein, and H.Morkoc: Improvement of the inverted GaAs/ AlGaAs heterointerface, J.Vac.Sci.Technol. B2 (1984)2,170-174

|26| Arnold,D., T.Henderson, J.Klem, R.Fischer, W.Kopp, A.Ketterson, W.T.Masselink, and H.Morkoc: High performance inverted and large current double interface modulation-doped field--effect transistors with the bulk (AlGa)As replaced by superlattice at the inverted interface, Appl.Phys.Lett. 45 (1984) 8,902-904

|27| Masselink,W.T., Y.L.Sun, R.Fischer, T.J.Drummond, Y.C.Chang, M.V.Klein, and H.Morkoc: Enhanced luminescence from AlGaAs/ GaAs single quantum well structures through improved interfaces, J.Vac.Sci.Technol. B2 (1984)2,117-122

|28| Masselink,W.T., M.V.Klein, Y.L.Sun, Y.C.Chang, R.Fischer, T.J.Drummond, and H.Morkoc: Improved GaAs/AlGaAs single quantum wells through the use of thin superlattice buffers, Appl. Phys.Lett. 44 (1984)4,435-437

|29| Drummond,T.J., J.Klem, D.Arnold, R.Fischer, R.E.Thorne, W.G. Lyons, and H.Morkoc: Use of a superlattice to enhance the interface properties between two bulk heterolayers, Appl. Phys.Lett. 42 (1983)7,615-617

Chapter 6. TRENDS IN RESEARCH AND TECHNOLOGY OF SUPERLATTICE STRUCTURES

In the history of semiconductor electronics, one notices that many important inventions and discoveries were made when basic physical concepts intersected with a new material technology or vice versa |1|. For example, the bipolar transistors was born, when the concept of "minority carrier injection and collection" crossed with the Ge p-n junction technology, whereas the semiconductor laser came into excistence, when the concept of "population inversion" or "stimulated emission" encountered the junction technology of direct-gap GaAs-related materials.

Another examples of such a creative interaction can be seen in the recent development of heterostructures with ultrathin layers. The fine epitaxy techniques, i.e. MBE and MO CVD, have enabled the preparation of various heterostructures, in which the quantum mechanical concepts of "size-quantization" and "tunneling" play an important role. By using such structures a number of new devices have been realized. Quantum well lasers |2| and high elec tron mobility transistors |3| may serve as good examples. To this group of heterostructures belong also superlattice structures. Reading the text of this book, the Reader should become familiar with the fact that the basic physical concepts relevant to different superlattices have one common feature. This is the vital role which the spatial coordinate z plays in generating the characteristic properties of the superlattices. The potential profile V(z), the wave-function $\psi(z)$, the charge distribution $\rho(z)$ and the average electron velocity v(z) are all position dependent, and this dependence is responsible for most of the superlattice properties that can be applied to new device concepts. This has generated a new field of device physics known as "wave-function engineering" |1|. This term makes a distinction from the more conventional device physics field known as "energy-band engineering" in which the manipulation of band structures in the $\vec{k}$-vector space is the main concern. Two new branches of "wave-function engineering" in which an intensive exploration can be foreseen for the nearest future have to be mentioned. One of such branches is the branch of quantum well wire structures |4| including planar or gate-controlled superlattice structures |5|, and the second is the branch of devices in which the wave function or localized charge carriers are controlled by external field in such a way that macroscopic quantities such as the conductivity or the light output can be modulated |6|. The velocity modulation transistor is one of the first examples of devices belonging to the second branch. The electron mobility in field effect transistor channels is changed by the gate field. Since the conductivity change in this device is not associated with the change of carrier density, such transistor is in principle free from the transit time limitations, thus being an ideal candidate for ultrafast electronic circuit applications.

In all these submicron structures heterointerfaces play a crucial role in defining the device parameters. However, devices in which superlattices are applied as active elements consist of large amount of interfaces. Therefore, the problem of the device quality may be in this case identified to a large extend with the problem of the performance of highest quality heterointerfaces.

As the development of devices with superlattices progresses, it calls for a constant interchange of ideas between the researchers acting in device physics or device technology and those involved in the more fundamentally-oriented surface/interface physics. This interchange goes both ways: on the one hand the device physics and technology can rely less and less on "cookbook empiricism", instead they both must closely follow the basic physics in assimilating and utilizing the new fundamental knowledge that the latter has acquired. On the other hand, device technology and physics constantly poses new problems demanding basic research. Experiments on device-type structures are themselves powerful tools for basic investigations |7|,|8|.

The recent development of basic research in surface and interface physics has resulted in a new physical picture of the interface region. Whereas in the mid-1960's an interface was regarded merely as the boundary between two bulk media, today it is seen as an independent entity: a state of matter determined by its own unique composition, structure, chemical bonding, electronic states and atomistic dynamics |9|.

The short inelastic collision mean free path of low-energy electrons (10 eV $\leqslant E_{kin} \leqslant$ 500 eV) rendered electrons when elastically emitted from the solid a direct probe of the few atomic layers nearest to the surface. This triggered the use of AES and XPS (X-ray photoelectron spectroscopy) for surface and interface analysis. The embodiment of this concept into theories of electron--solid scattering led to modern models of low-energy electron diffraction (LEED), which, in turn, have yielded the vast majority of surface structure determinations. The atomic geometries of single-crystal surfaces and interfaces became, thus, directly accessible to experimental determination. The results of these structure determinations have been enlightening and, in many cases, unexpected. For example, the clean (110) surfaces of zincblende structure compound semiconductors reconstruct to form structures intermediate between molecular and crystal geometries for III-V semiconductors. Remarkably, the same surface structures persist for II-VI compounds, correlating with bulk covalent radii rather than bonding ionicities or small-molecule geometries |9|. The theme illustrated by these examples is self-evident: observed surface structures often differ from those which were expected "a priori". Thus, the emergence of quantitative surface and interface structure analysis has led to a new view on interfaces, once their actual (as opposed to hypothetical) structures became known

Improvements in quantitative surface structure analysis have been paralleled by those in the calculation of the electronic properties of surfaces and interfaces. A major motivation for the construction of realistic models related to the electronic structure of the interfaces has been the development of spectroscopic techniques which can measure directly special features of this structure. It was recognized that any type of surface or interface electronic state can be observed via valence electron photoemission. Thus, both the experimental measurements and their interpretations constitute examples of the ongoing refinement of the interface-as-an-entity paradigm introduced by the surface physicists.

There are two important aspects controlling the properties of semiconductor superlattices, namely, the interface atomic geometry, and the energy band, or crystal potential line-up, at the interfaces. These two basic ingredients are not independent of each other, but rather an intimate interdependence is expected on theoretical grounds and experimental studies |10|.

The quality of the semiconductor superlattices depends, however, mainly on the perfection, which can be reached in crystal growth. An interesting example of the attempts in this direction has been reported by Sakamoto et al. |10|. This is the so-called "phase--locked epitaxy" (PLE) based on the intensity oscillations observed in the reflection high energy electron diffraction (RHEED) patterns during the MBE growth of GaAs and AlGaAs layers. The apparatus used in the PLE is schematically shown in Fig. 6.1.

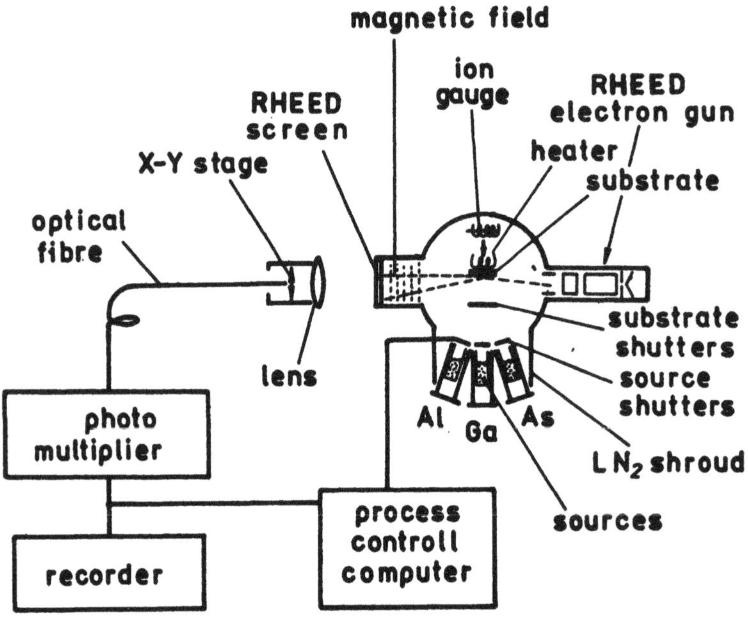

Fig.6.1. Schematic diagram of the phase-locked epitaxy system using RHEED intensity oscillations (after Sakamoto et al., Ref.10)

The idea of the method is as follows. Long continuing intensity oscillations of the RHEED pattern in the [100] azimuth on a (001) oriented substrate were observed during MBE growth of GaAs and the

$Al_xGa_{1-x}As$ (see, Fig.6.2a). Using these oscillations growth rates of GaAs and $Al_xGa_{1-x}As$, and the Al mole fraction x of the ternary compound were accurately monitored during the growth. The phase of the RHEED oscillations was analysed by computer and molecular beam shutters were operated at a particular phase of the oscillations.

This computer controlled PLE was used to grow precisely defined $(GaAs)_2(AlAs)_2$ bi-layer superlattices. Raman scattering spectra of the bi-layer superlattice showed split lines characteristic of

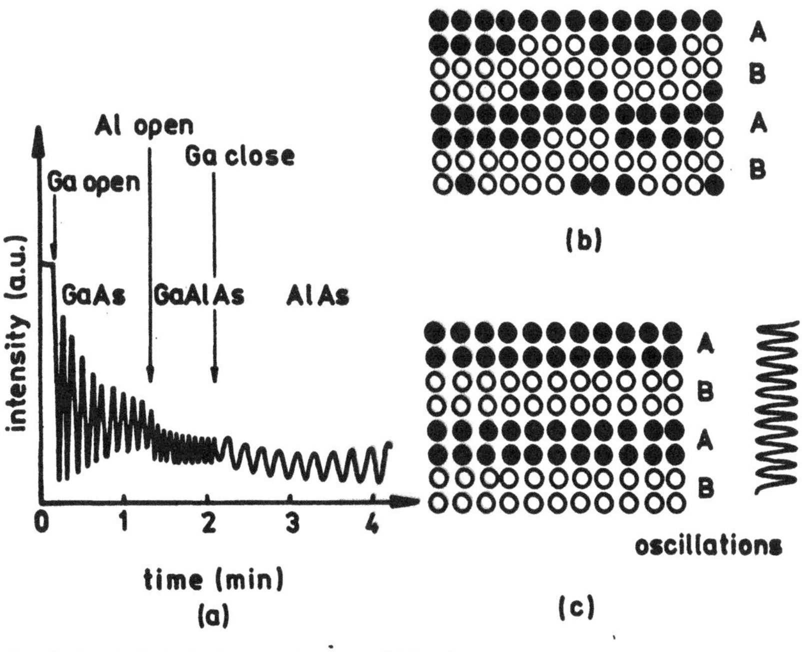

Fig.6.2.a) RHEED intensity oscillations of the specular beam observed in [100] azimuth of (001) GaAs substrate during the continuous growth of GaAs,$Al_xGa_{1-x}As$ and AlAs.
b) and c) Simplified models of the cross sectional view of superlattice structures grown by two methods: the conventional non-PLE method (b), and the PLE method (c).
(after Sakamoto et al.,Ref.10)

superlattices. From TEM observation of a GaAs-AlAs multi-layered structure, it was verified that one cycle of oscillations corresponds to one monolayer growth of GaAs and AlAs. This PLE has a great advantage over the conventional MBE growth method for the precise control of very thin films and superlattices because it is invulnerable to fluctuations of molecular beam flux intensity. Simplified models of the cross-sectional view of superlattice structures grown by these two methods are shown in Fig. 6.2. In the case of the non-PLE method (see,Fig. 6.2b) molecular beam shutters are operated at a random phase of the surface molecular coverage and as a result, the heterostructure interfaces between A and B have steps. On the other hand in the case of the PLE method, the interfaces between A and B become very flat, as shown in Fig. 6.2c. since the molecular beam shutters are operated at the maximum amplitude of the oscillations when the monolayer is completed. A $(GaAs)_2(AlAs)_2$ bi-layer superlattice with 699 periods, i.e., with 2796 monolayers was grown using this computer-controlled PLE method, with excellent structural quality |10|.

A new dimension to the application area of semiconductor superlattices could be probably added by introducing the one-dimensional organic crystal structures into the technology of devices with superlattices.

Wires that are molecular chains are being recently developed |11-13|. These wires are crystals of trans-poly-di-acetylene (PDATS), an organic polymer. In most polymers the molecules are randomly arranged coils of molecular chains. But in crystals of PDATS, the molecular chains are in long straight lines. When shining short pulses (10 ns) of laser light at the crystal of PDATS free electrons are created |11|. These electrons move in an applied electric field, with a speed of about $2.2\cdot10^3$ ms^{-1} |13|, which is fairly constant regardless of the size of the field applied. The magnitude of the low-field mobility is ultra-high, exceeding the value of $2\cdot10^5$ $cm^2V^{-1}s^{-1}$. This is a result of the motion of a solitary wave (soliton) acoustic polaron in one dimension |12|,|14|. Although the ideas opened up by this investigations are exciting, practical application of these one-dimensional wires in devices with superlattices seems to be a fairly long way off.

Concluding this considerations on trends in research and technology of SL structures, it should be emphasized that only a close interaction between the basic research on crystal growth, as well as on structural and electronic properties of superlattices, with the device electronics, can lead to the full application of the enormous potential which the variety of superlattice structures offer for the electronic praxis.

6.1. Literature to chapter 6

|1| Sakaki,H.: Evolution of new device concepts in quantum well and superlattice structures, Extended Abstracts of the 15-th Conf.on Solid State Dev. and Mater., Tokyo, 1983, pp.3-6

|2| Holonyak,Jr,M.: Quantum well and superlattice lasers-fundamental effects, in: The Physics of Submicron Structures, H.L. Grubin, K.Hess, G.J.Iafrate, and D.K.Ferry (eds.), Plenum Press, New York, 1984, pp.1-18

|3| Morkoc,H.: The HEMT-a superfast transistor, IEEE Spectrum, Febr. 1984, pp.28-35

|4| Petroff,P.M., A.C.Gossard, R.A.Logan, and W.Wiegmann: Toward quantum well wires-fabrication and optical properties, Appl. Phys.Lett. 41 (1982)7,635-638

|5| Stiles,P.J.: The artificial superlattice in a two-dimensional system, Surface Sci. 73 (1978)1,252-257

|6| Hirakawa,K., H.Sakaki, and J.Yoshino: Mobility modulation of the 2DEG via controlled deformation of the electron wave function in selectively doped AlGaAs/GaAs heterojunctions, Phys. Rev.Lett. 54 (1985)12,1279-1282

|7| Kroemer,H.: Heterostructure devices-a device physicist look at interfaces, Surface Sci. 132 (1983)1,543-567

|8| Margaritondo,G.: The heterojunction parameters from a microscopic point of view, Surface Sci. 132 (1983)1,468-478

|9| Duke,C.B.: Atoms and electrons at surfaces-a modern scientific revolution, J.Vac.Sci.Technol. A2 (1984)2,139-143

|10| Sakamoto,T., H.Funabashi, K.Ohta, T.Nakagawa, N.J.Kawai, T. Kojima, and Y.Bando: Well defined superlattice structures made by phase-locked epitaxy using RHEED intensity oscillations, Superlattices and Microstructures 1 (1985)4,347-352

|11| Crystal chains for electronics, New Scientist 106 (1985) 1458, 23

|12| Donovan,K.J., P.D.Freeman, and E.G.Wilson, Acoustic solitary wave polaron motion in polydiacetylene crystals, Materials Sci. 10 (1984)1/2,61-66

|13| Donovan,K.J., and E.G.Wilson, Carrier trapping in one-dimensional systems: example organic PDATS, J.Phys.C: Solid State Phys. 18 (1985)2,L51-L54

|14| Lonngren,K., and A.Scott eds., Solitons in Action, Academic Press, New York 1978

PRINCIPAL SYMBOLS

a_o - crystal lattice constant(lattice parameter)
d - layer thickness, density of a material
d_i - thickness of the intrinsic layers of a doping SL
$d_n(d_p)$ - thickness of the n-type (p-type) layers of a doping SL
e - absolute value of the elementary charge
$\hbar$ - Planck's constant h divided by 2π
k - Boltzmann's constant
$\vec{k}$ - wave vector of a particle
m - mass of a free particle
m_o - mass of a free electron
m_e^* - electron effective mass in a semiconductor
m_h^* - hole effective mass in a semiconductor
n - electron concentration, refractive index
p - hole concentration, value of the particle momentum, pressure
$\vec{r}$ - radius-vector of a material point
t - time
u - Bloch's wave function
x,y - Cartesian co-ordinates in the direction parallel to the potential well walls, indices defining the composition of a compound
z - Cartesian co-ordinate in the direction of the SL axis
E - energy, eigenvalues of the Hamiltonian
$E_A(E_D)$ - ionization energy of an acceptor (donor) in a semiconductor
E_C - conduction band-edge, the potential energy of an electron in semiconductor lattice
$E_{C,i}$ - energy of the i-th SL subband in the conduction band
E_F - Fermi energy
E_{F_n}, E_{F_p} - quasi-Fermi levels in n- and p-type semiconductors, respectively
E_g - energy gap of a bulk semiconductor
$E_{g_{SL}}$ - SL effective band gap
E_i - impurity binding energy
E_V - valence band-edge, the potential energy of a hole in semiconductor lattice
$E_{V,i}$ - energy of the i-th SL subband in the valence band

ΔE_C - discontinuity of the conduction band-edges in a heterojunction

ΔE_g - discontinuity of the energy gap in a heterojunction

ΔE_V - discontinuity of the valence band-edges in a heterojunction

F - Bloch's envelope function, structure factor in X-ray diffraction

F_z - strength of the electric field in the direction of the SL axis

H - Hamiltonian

L - thickness of a layer

L_z - thickness of a quantum well

N_A - concentration of acceptors in a semiconductor, Avogadro's number

N_D - concentration of donors in a semiconductor

P_{mn} - matrix element of the carrier momentum

Q_i - fraction of ΔE_g in the i-th energy band

R - gas constant

T - absolute temperature

V_D - space-charge (built-in) potential

V_{D_o} - amplitude of the periodic space-charge potential of a doping SL

X-rays - Roentgen radiation

α - optical absorption coefficient, impact ionization rate for electrons

α_A - thermal accommodation coefficient

α_S - sticking coefficient

β - impact ionization rate for holes

δ - Dirac's delta function

δ_{nm} - Kronecker's delta function

ε - dielectric constant

$\hat{\varepsilon}$ - strain tensor

θ - electron affinity

μ - molecular weight

ψ - quantum-mechanical wave function

$\vec{\xi}$ - vector of the electric field strength

APD - avalanche photodiode

HWE - hot wall epitaxy

LED - light emitting diode

LEED - low energy electron diffraction

LPE - liquid phase epitaxy
MBE - molecular beam epitaxy
MO CVD - metalorganic chemical vapour deposition
MQW - multi-quantum well
PLE - phase-locked epitaxy
RHEED - reflection high energy electron diffraction
SL - superlattice
SLS - strained-layer superlattice
TEM - transmission electron microscopy
2DEG - two dimensional electron gas
2DHG - two dimensional hole gas

INDEX

P H Y S I C A L   R E S E A R C H

Akademie-Verlag Berlin is beginning a complementary series to its range on Mathematical Research in 1985. The new series, Physical Research, follows the international trend of publishing short monographs on current research results as quickly as possible. Subjects from all fields relevant to international physical research are to be dealt with, predominately in English and in manuscript form.

In preparation:

Harald Böttger/Valerij V. Bryksin
Hopping Conduction in Solids

Klaus Baumgärtel/Konrad Sauer
Topics on Nonlinear Wave Plasma Interaction

Helmut Heß
High-Pressure Plasmas

Eckehardt W. Mielke
Geometrodynamics of Gauge Fields

Manfred Hoffmeister
Methods of Hot-Wire Anemometry

Friedhelm Bechstedt/Rolf Enderlein
Semiconductor Surfaces

P a r c e l l a '84

Proceedings of the II. International Workshop
on Parallel Processing by Cellular Automata and Arrays
held in Berlin (GDR), September 25 - 27, 1984

---

edited by

Wolfgang Händler

Tamás Legendi

Gottfried Wolf

This volume contains the papers of PARCELLA '84 - the II. International Workshop "Parallel Processing by Cellular Automata and Arrays" held in Berlin from September 25 to 27, 1984.

This workshop united those scientists who were interested in discussing the new results and ideas for parallel and distributed processing on homogeneous computer structures. Cellular systems and array systems are main representatives of this structure type.

The existence side-by-side and the competition between these two closely related concepts appears to be fruitful for their own development as well as for the overall development of further computer architecture in general.

Bei Fragen zur Produktsicherheit wenden Sie sich bitte an:
If you have any questions regarding product safety,
please contact:

Walter de Gruyter GmbH
Genthiner Straße 13
10785 Berlin
productsafety@degruyterbrill.com